738·134

Frankhauser W.

ADVANCED PROCESSING OF
CERAMIC COMPOUNDS

ADVANCED PROCESSING OF CERAMIC COMPOUNDS

Dynamic Compaction Technology, Self-Propagating High-Temperature Synthesis, Plasmachemical Technology

by

William L. Frankhouser

System Planning Corporation

NOYES DATA CORPORATION

Park Ridge, New Jersey, U.S.A.

1987

Sole distribution by:
Gothard House Publications
Gothard House, Henley-on-Thames,
Oxon RG9 1AJ
Tel: 0491 573602

Copyright © 1987 by Noyes Data Corporation
Library of Congress Catalog Card Number 87-12252
ISBN: 0-8155-1130-2
Printed in the United States

Published in the United States of America by
Noyes Data Corporation
Mill Road, Park Ridge, New Jersey 07656

10 9 8 7 6 5 4 3 2 1

Library of Congress Cataloging-in-Publication Data

Frankhouser, William L.
    Advanced processing of ceramic compounds.

    Bibliography: p.
    Includes index.
    1. Ceramics. I. Title.
TP807.F73 1987   666   87-12252
ISBN 0-8155-1130-2

Foreword

This book describes both established and potential technology for the processing of advanced ceramic compounds, materials which would satisfy key performance functions and properties in aerospace and defense applications and, subsequently, in the industrial sector. Particularly covered are dynamic compaction technology, self-propagating high-temperature synthesis, and plasma-chemical technology, the first two of which are on the leading edge of materials technology, and, when fully investigated, are expected to be of great value to commercial interests. The book covers research in the U.S. and the U.S.S.R. The future increase in performance requirements of unique materials underlies the importance of these specialized processing techniques.

Rapid market growth during the next ten to fifteen years has been predicted for products fabricated from advanced ceramic compounds. Examples of the wide range of products which might be fabricated by the processes discussed include: high temperature engine components such as combustion liners, blades, nozzles, and shrouds; electronic hardware such as thermistors, insulators, and substrates for integrated circuits; other refractory structurals such as furnace elements, heat-exchanger components, and metal-refining electrodes; energy storage devices; materials-finishing accessories such as grinding wheels, polishing pastes, and machining tool bits; and wear and abrasion hardware.

Specific applications of processing technologies in government programs are presented in the book, and representative materials compositions and manufacturing procedures are briefly described. Sample assessments of the commercial potential for several ceramic products are also provided.

The information in the book is from *Advanced Materials Technology Project—Final Technical Report,* prepared by William L. Frankhouser of System Planning Corporation for the U.S. Department of Defense, Defense Advanced Research Projects Agency, May 1986.

The table of contents is organized in such a way as to serve as a subject index and provides easy access to the information contained in the book.

Advanced composition and production methods developed by Noyes Data Corporation are employed to bring this durably bound book to you in a minimum of time. Special techniques are used to close the gap between "manuscript" and "completed book." In order to keep the price of the book to a reasonable level, it has been partially reproduced by photo-offset directly from the original report and the cost saving passed on to the reader. Due to this method of publishing, certain portions of the book may be less legible than desired.

Acknowledgments

The author wishes to acknowledge and extend thanks for the effective efforts of others who have participated in the total program that is reported herein:

Capable technical support was provided by Keith Brendley, Michael Kieszek, and Stephen Sullivan of SPC, and by Francis Quigley, an SPC consultant.

The great number of Soviet translations were performed by Judith Gogolewski of SPC.

The research was ably assisted by Linda Glickman, Nicholas Mercury, Phyllis Moon, and other members of SPC's Technical Library staff.

The text was edited by Len Eggert and Ellen Levenseller. Typing was provided by Doreen Kelley and Janice Taylor, who are extremely tired of spelling Soviet names.

The many reports prepared under this effort were illustrated by SPC's Visual Department and reproduced by the Publications Center.

Guidance from the Program Manager at DARPA, Dr. Steven Wax, and from SPC management, specifically Jack Fagan and John Drain, helped immeasurably.

Overall, the program has been a gratifying experience. Although the investigative subjects are on the leading edge of materials technology, practical applications are now in sight.

NOTICE

The materials in this book were prepared as accounts of work sponsored by the U.S. Department of Defense, Defense Advanced Research Projects Agency. The views, opinions, and findings contained in this report are those of the author and should not be construed as an official Department of Defense position, policy, or decision, unless so designated by other official documentation. On this basis the Publisher assumes no responsibility nor liability for errors or any consequences arising from the use of the information contained herein. Mention of trade names or commercial products does not constitute endorsement or recommendation for use by the Agency or the Publisher.

Final determination of the suitability of any information, procedure, or product for use contemplated by any user, and the manner of that use, is the sole responsibility of the user. The book is intended for informational purposes only. Expert advice should be obtained at all times when implementation is being considered, particularly where hazardous materials or processes are encountered.

Contents and Subject Index

I. Executive Summary

A. **PROJECT DESCRIPTION AND ORGANIZATION OF REPORT**

A final technical report on System Planning Corporation's (SPC) Advanced Materials Technology Project has been completed for the Defense Advanced Research Projects Agency (DARPA). This report integrates information presented previously in quarterly progress reports and semiannual technical reports with information on activities performed during the present quarter of the 2-year contractual period.

Laboratory development of advanced materials processing technology has been sponsored by MSD at Battelle (BTL), Lawrence Livermore National Laboratory (LLNL), and Los Alamos National Laboratory (LANL). Emphasis in laboratory programs has been on fabrication of advanced refractory ceramic compounds by new technologies that promise quality products not otherwise available at reasonable cost. The processing techniques are dynamic compaction technology (DCT), self-propagating high-temperature synthesis (SHS), and plasmachemical technology (PCT).

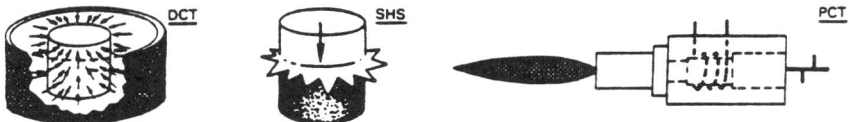

SPC has supported both MSD and the laboratory participants through investigations of commercial potential of the processing technologies and opportunities for technology transfer and by assessments of Soviet progress in materials science and engineering. The four specific task assignments are:

1

- Technology applications in commercial industry (Task 1)
- Technology applications within the defense community (Task 2)
- Relevant Soviet materials technology (Task 3)
- Commercial potential of technology (Task 4).

Activity highlights for the four tasks are presented in Section C of this chapter. More detailed accounts of activities follow in Chapters II through V, where locations of key information in prior SPC reports also are identified.

B.   SIGNIFICANCE OF ADVANCED CERAMICS AND MSD FABRICATION TECHNOLOGIES

Implications attributed to top scientists [Refs. 1 through 13] regarding emergence of major scientific developments during the remainder of this century follow:

- Performance functions requiring unique materials properties underlie emergence of the most significant developmental technologies into new industrial products.
- Many property requirements reach beyond performance capabilities of existing materials.
- Space and energy systems that operate at high levels of efficiency, at extreme temperatures, and in harsh environments dominate development of new materials technologies.
- Most new products that satsify these needs must be fabricated through powder processing operations, and many of these products are advanced ceramics.

As demonstrated in Figure 1, the DARPA development of advanced processing technologies is an appropriate response to crucial needs of emergent scientific developments. The program concentrates on processing advanced ceramic materials that will satisify many key performance functions and properties.

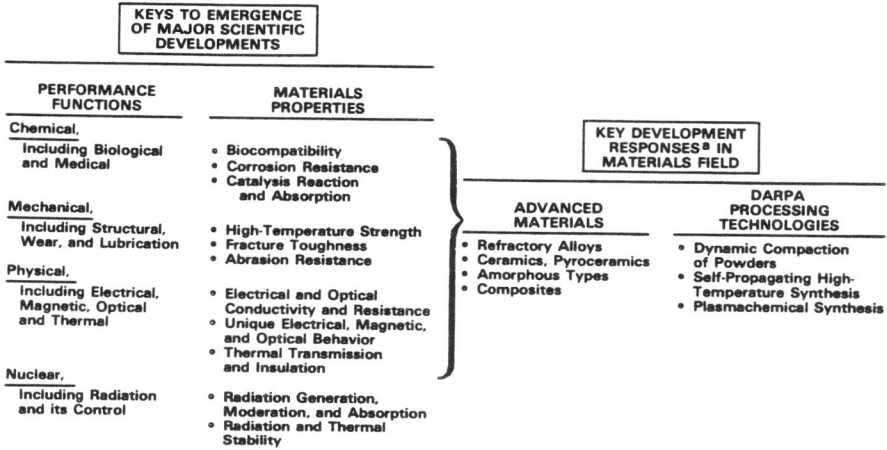

KEYS TO EMERGENCE OF MAJOR SCIENTIFIC DEVELOPMENTS

PERFORMANCE FUNCTIONS	MATERIALS PROPERTIES
Chemical, Including Biological and Medical	• Biocompatibility • Corrosion Resistance • Catalysis Reaction and Absorption
Mechanical, Including Structural, Wear, and Lubrication	• High-Temperature Strength • Fracture Toughness • Abrasion Resistance
Physical, Including Electrical, Magnetic, Optical and Thermal	• Electrical and Optical Conductivity and Resistance • Unique Electrical, Magnetic, and Optical Behavior • Thermal Transmission and Insulation
Nuclear, Including Radiation and its Control	• Radiation Generation, Moderation, and Absorption • Radiation and Thermal Stability

KEY DEVELOPMENT RESPONSES[a] IN MATERIALS FIELD

ADVANCED MATERIALS	DARPA PROCESSING TECHNOLOGIES
• Refractory Alloys • Ceramics, Pyroceramics • Amorphous Types • Composites	• Dynamic Compaction of Powders • Self-Propagating High-Temperature Synthesis • Plasmachemical Synthesis

[a]WIITH RELEVANCE TO DARPA PROGRAM

FIGURE 1. SIGNIFICANCE OF THE DARPA MATERIALS PROCESSING PROGRAM TO MAJOR SCIENTIFIC DEVELOPMENTS

C.   STUDY HIGHLIGHTS

1.   Technology Transfer to Commercial Industry (Chapter II)

The task assignment was to facilitate technology transfer to commercial industry. The approach was to make industry aware of the DARPA program and to encourage flow of information between the laboratory participants and interested industrial organizations. The effort was centered on fabricating advanced ceramics and composites by SHS and DCT technologies, with greater emphasis on the former; PCT development was not considered to be sufficiently mature to warrant a comprehensive effort.

The largest volume market applications anticipated for advanced ceramic and composite products in six industrial fields are identified in Table 1. Each product is associated either directly with specific needs of major emergent technologies or indirectly through support industries.

In regard to SHS technology, the interest in commercial industry is judged to be intense:

- Thirty-five points of contact in industry have been briefed individually on SHS.

- Fifteen to 20 industrial companies were represented at the program review that was held in Florida during October 1985.

- A book describing U.S. and U.S.S.R. programs in SHS has been published [Ref. 14].

- SHS was the subject of a feature article in Advanced Materials & Processes during February 1986 [Ref. 15].

- The technology has been described to a sizeable group of industrial representatives in a 2-day seminar on advanced materials processing technologies.

- An industrial consortium is being organized to support continued technology development at LLNL as financial support is phased out by DARPA

TABLE 1

**ADVANCED INDUSTRIAL PRODUCTS POTENTIALLY FABRICABLE
BY DARPA PROCESSING TECHNOLOGIES**

HIGH-TEMPERATURE ENGINES

- Combustion engines (e.g., combustion liners, chambers)
- Turbine engines (e.g., blades, nozzles, shrouds)

ELECTRONIC HARDWARE

- Thermistors, semiconductors
- Insulators
- Substrates for integrated circuits

OTHER REFRACTORY STRUCTURALS

- Furnance elements
- Heat exchanger components
- Heat pipes
- Metal refining electrodes

ENERGY STORAGE

- Advanced electromagnetic batteries
- Hydrogen storage systems

MATERIALS FINISHING ACCESSORIES

- Grinding wheels, polishing pastes
- Machining tool bits

WEAR AND ABRASION HARDWARE

- Metal-forming dies
- Pump components (e.g., seals)

Although the DCT program was not briefed as widely to commercial industry as SHS technology, considerable industrial interest has been apparent. An effort has been initiated recently to present a seminar on DCT to a large group of industrial representatives. Also, many industrial products that can be fabricated by dynamic compaction of metal or ceramic powders have been identified to representatives of the laboratory development programs.

The PCT program at LANL was described in an independent technology transfer briefing at that facility. This technology was one among many LANL developments described to industry.

2.   Potential Defense Systems Applications (Chapter III)

The task assignment was to establish program interfaces for eventual defense applications of the DARPA materials processing technologies. The approach was to identify potential applications and to develop appropriate contacts between the materials processing program and Government systems programs. The effort was centered on DCT and SHS technologies, with greater emphasis on the former. PCT development was not considered to be sufficiently mature to warrant a comprehensive effort.

Since advanced ceramics are important to future defense applications and other Government development programs, a comprehensive assessment was made of the expected evolution of an advanced ceramics industry during the remainder of this century. Results indicate that:

- The industry is expected to demonstrate one of the most rapid market growths during the 1980s and 1990s among all industries in developed countries.

- Many of the potential applications are attributable to processing capability in synthesizing compounds, compacting powders, or coating substrates.

- The greatest potential processing payoff is combined synthesis and compaction of dense products.

- Rapid solidification technology (RST) and DCT make an ideal processing combination by which new amorphous materials (e.g, metallic glasses and complex ceramic compositions) can be formed as powders and subsequently densified into monolithic product shapes without undesirable crystallization.

Specific applications for DARPA processing technologies in Government programs are presented in Table 2 for six product types, and representative material compositions (e.g., aluminides, nitride/carbide and metal/ceramic composites, hydrides) and manufacturing procedures are described in Chapter III. The capability of DCT and SHS technology in bonding macrocomposite structures and in making microcomposite compositions is exploited in these applications.

Points of contact were established between the DARPA advanced processing technology program and other Government programs:

- Approximately 25 program managers were provided with briefing information on the DCT and SHS technology programs.

- Seven of these managers attended the DCT program review in September 1985, and nine Government organizations, in addition to MSD, were represented at the SHS program review in October 1985.

- A follow-on program to continue development of DCT in the armaments field when DARPA funding is phased out has been planned by these contacts.

3. Surveillance of Relevant Soviet Programs (Chapter IV)

The task assignment was to assess relevant Soviet work in developing and processing advanced materials and to assess their product applications. The approach was to monitor open (unclassified) Soviet literature on materials science and engineering. In cases where commercial translations of Soviet publications into English were not available, translations were made and forwarded to all participants in the DARPA program. Greatest emphasis was on PCT since this was considered to be in the earliest development stage among the three processing technologies.

Soviet development of PCT and SHS continues to lead progress in the United States:

- Both technologies have been included as discrete items for several years in the Soviet national plan for advancement of technology.

- National Soviet centers have been established to exploit each technology.

TABLE 2

POTENTIAL PRODUCT APPLICATIONS IN GOVERNMENT PROGRAMS[a]

Applications	Potential Products Fabricable by DARPA Processing Technologies
Armaments	Lightweight ceramic armor Ceramic gun barrels or liners Composite (metal/ceramic) penetrators
Ceramic Engines	Combustion components and hardware Spark plugs Turbine blades Bearings
Electronic and Optical Hardware	Stacked-layer capacitors and semiconductors Fiber optic wave guides on ceramic substrates
Other Structural Components	Composite (metal/ceramic) periscope tubes Deep-submergence vessels Space reactor heat pipes Memory-alloy connectors
Nuclear Power	Advanced fuels Waste isolation matrices High-efficiency neutron shields Refractory alloys and nonmetallic compounds (for heat pipes, structurals, liquid metal service, etc.)
Mechanical	Tool hardware High-temperature lubricants

[a]Some of the broader industrial applications previously listed in Table 1, although applicable, are not repeated.

- Both technologies are taught in college-level curricula in the U.S.S.R.

- Lasers have been used by Soviet scientists to carbidize and nitride metal surfaces from plasmas.

Industrial applications of both PCT and SHS technology are reported in the U.S.S.R., with claims of significant benefits to the national economy:

- SHS products include high-temperature heating elements (molybdenum disilicide); also, tool bits, drilling crowns, grinding wheels, and polishing pastes are made from titanium carbide or carbonitride compounds.

- SHS titanium carbide and carbonitride compounds are rapidly replacing compounds containing the strategic metal tungsten.

- The Bureau of Tank Production recently assumed responsibility for the SHS program--presumably, to take advantage of improved machine tool products.

- Essentially all titanium dioxide paint pigment is produced by PCT.

- Metals (e.g., molybdenum alloys and steel) are coated with ceramic compounds by PCT in order to extend service life through increased resistance to high-temperature oxidation or to wear and abrasion.

- Many ceramic powders (e.g., nitrides and carbides) that are difficult to produce, or not producible at reasonable cost, by alternative technologies have been prepared by PCT.

A major difference in emphasis on development of DCT has been noted in the United States and U.S.S.R.:

- The DARPA program emphasizes simple compaction of powder to produce monolithic shapes.

- The Soviet program has emphasized phase transformation and synthesis. [However, a 1985 paper describing powder compaction of a cermet product may be the harbinger of another important Soviet technology application.]

A recent assessment of Soviet approaches to increasing service temperature and improving overall performance of refractory structural materials resulted in the following conclusions:

- Complex compositions of Mo, Nb, Ta, V, and W alloys, which combine solid solution and dispersion particle strengthening mechanisms, are being tested.

- Surface coatings on refractory metal alloys, primarily aluminide and silicide, are used for further extension of service temperatures.

- Complex ceramic and composite compositions with delocalized electronic structures, often based on elemental additions to TiC or B_4C, have demonstrated significant gains in fracture toughness over binary ceramic compositions.

Although U.S. literature in materials science is used widely in Soviet R&D programs, open Soviet literature has not been exploited at a comparable level in the United States. A U.S. clearinghouse is needed to assess (and translate, when necessary) Soviet writings on technology in materials science and engineering and to distribute the information (unclassified) widely to industry, academia, and R&D laboratories where it can be utilized most effectively.

SPC translations of Soviet documents are identified in Chapter IV. These documents constitute a valuable source of information on DCT, SHS, and PCT.

4.   Assessment of Commercial Potential (Chapter V)

The assignment was to assess commercial potential of processing technologies. The approach was to estimate production-scale prices for specific ceramic products that were fabricated by either DCT or SHS technology and to compare [them] with contemporary market prices.

A base price was calculated for initial production by DCT of cylindrical ceramic disks that are encased in a thin metal jacket and would be used in metal/ceramic assemblies. This price was taken as the starting value for predictions of product price over an extended period wherein production quality increases steadily:

- The base price of $81 per pound of contained ceramic compound is somewhat lower than recent selling prices for smaller hot-pressed bare ceramic tiles of the same composition.

- Over an 11-year period of production, the base price was projected to decrease to a level between $8 and $39 per pound (with a most likely projection of $12 per pound).

A similar assessment was made for manufacture of bare ceramic tiles by SHS technology:

- The base price is $51 per pound of contained ceramic compound.

- After the 11th production year, the most likely price projection is $8 per pound.

- The lower values for SHS processing, in comparison to DCT, are attributed mainly to lower cost of ceramic raw material and to process simplification, where synthesis and densification are combined in a single-step operation.

Other significant conclusions from these price projections for the ceramic products follow:

- The base price is influenced heavily by the cost of ceramic starting powders (equalling approximately one-half of the total cost).

- Competitive influence in the market coupled with production learning experiences will drive the price downward in a market that is characterized by rising demand.

- The rate of price decline will not be appreciably different for any advanced processing technology when market demand increases steadily.

II. Task 1: Technology Applications in Commercial Industry

A. POTENTIAL INDUSTRIAL PRODUCTS

The initial activity in this task was a review of emergent technological developments anticipated during the remainder of this century. This review identified not only the more significant advanced developments but also the greatest needs in regard to performance requirements of materials and their properties (Figure 1). The industrial products listed in Table 3 comprise many items that will be manufactured by commercial industry as a result of these technological advances.

Emphasis in this list is on unique electronic, mechanical, and refractory properties of materials. Prominent in the list are advanced ceramics, composites, and the amorphous (or near-amorphous) metallic glasses, plus some metals and alloys. Ceramics often promise better high-temperature strength and strength/weight efficiency than metals, but the problem of fracture toughness must be solved. Ceramic or intermetallic materials may be key in perhaps the most significant technological breakthrough in the list--achievement of electrical superconductivity at temperatures of liquid nitrogen or above. Directional solidification in metals, alloys, and intermetallics and production of metallic glasses in net product shapes also are expected to contribute significantly to this evolution of advanced technologies.

B. COMMERCIAL INDUSTRY INTERFACES

Initial contacts with commercial industry were made mostly by telephone in exploring interest in DARPA's materials processing programs in SHS and DCT. With a few exceptions, all individuals contacted expressed interest in SHS; however, DCT attracted only limited interest. These telephone calls were followed up with mailing packages of briefing information.

11

TABLE 3

**POTENTIAL INDUSTRIAL PRODUCTS
IN EMERGENT TECHNOLGICAL FIELDS**

Electronic Components

- Superconductors (niobium compounds)

- Thermistors, magnets (cobalt and nickel ferrites, $SmCo_5$)

- Temperature-independent electrical resistors (e.g., NbN, TaN, TiN, Ti_xV_yN, VN)

- Insulators and substrates for circuitry (e.g, AlN, Al_2O, beryllium oxide, magnesium aluminate, strontium and barium silicates)

- Memory metals (e.g., titanium nickelide)

- Recording heads, memory devices (magnesium and manganese ferrites)

- Semiconductors (e.g., B_xC/LaS_y, rare earth thiohafnates, rare earth thiozirconates)

- Dielectric transducers (barium and lead titanates and zirconates)

Energy Storage

- Batteries (e.g., with $SmCo_5$)

- Hydrides (e.g., $ZrCoH_3$, $ZrNiH_3$, MgH_2, NiH_2)

Materials Finishing Products

- Abrasives and grinding materials

  -- Bonded disks and wheels (e.g., using SiC, TiC, Ti(C,N), Ti_3AlC)

- Machining aids

  -- Ceramic and cemented carbide tool bits (e.g., using TiC, Ti(C,N), $(Ti,Cr)B_2$, (Ti,Ta)C, Si_3N_4, TiC with Mo/Re binder, WC with Co binder, W_2C, diamond, Al_2O_3, CBN)

  -- Surface coating of tools (e.g., using Si_3N_4, TiC with Mo/Re binder, WC with Co binder, W_2C)

- Polishing Compounds

  -- Grits and pastes (e.g., using TiC, Ti(C,N), Ti_3AlC)

TABLE 3 (Continued)

Refractory Products

- Adiabatic engine/components,[a] rocket nozzles (e.g, AlN, Cr_7C_3, SiC, Si_3N_4, TiC, ZrB_2)

- Dispersion-strengthened metals (e.g., Al_2O_3 in Al, HfB_2 in Mo or W, TiB_2 or TiS_2 in Ti)

- Furnace, elements, heat exchangers, heat pipes (e.g., $LaCrO_3$ + Cr, $MoSi_2$, $YCrO_3$ + Cr, SiC)

- Metal refining electrodes (e.g., Cu_2Al, NbB_2, TiB_2)

- Protective cladding (e.g., $AlCr_{23}C_6$ on Ni-base superalloys, CrB_2 or TiB_2 on steel, NbB_2 or $MoSi_2$ on Nb alloys, nickel aluminides or silicides on stainless steel or graphite, bonded macrolaminations of AlN and $MoSi_2$)

- Structural microcomposites (e.g., $LaCrO_3$ + Cr, C (fiber) in TiC, WC + Al_2O_3, ZrB_2 + Al_2O_3, Ti(C,N) with gradated composition)

Wear and Abrasion Products

- Dies for drawing, forming, and extrusion (e.g., transition metal carbide cermets, TiN)

- Surface coating (e.g., ZrB_2 on stainless steel)

---

[a]For example, gas turbine shrouds, piston caps, turbocharger rotors, valve heads and seats, manifold liners.

The briefing on SHS technology, which was prepared by SPC in cooperation with DARPA, is presented in Appendix A. The presentation describes SHS processing and its anticipated sphere of influence in the materials industry. The DARPA program laboratory participants are listed, program objectives are stated, and diagnostic data from some experiments are included. Products produced by SHS technology, both within the DARPA program and in the Soviet Union, are identified; also, potential industrial applications in the United States are discussed.

Corporations contacted regarding in SHS technology are identified in Table 4. Since several persons within this list asked to attend the review of the national U.S. program that was to be held later in Daytona Beach, Florida during 21-23 October 1985, a number of invitations were issued to industry by the review coordinator.[1] Approximately 15 industrial companies sent representatives to the review, including approximately 10 from the list in Table 4.

One of the representatives of a large U.S. corporation attending the review reported purchase of 40 pounds of silicon nitride powder that was produced by SHS technology in the Soviet Union. His subsequent character-ization of the material revealed that it was equivalent in quality to sili-con nitride powder produced elsewhere (e.g., in Japan) by more conventional technology.

The attendees at the review also included a representative from **Advanced Materials & Processes** magazine. An article on SHS technology was published subsequently in the February 1986 issue of the magazine [Ref. 15]. The article was based on information obtained at the review and from additional documents supplied by SPC and others.

An SPC report on SHS technology was modified slightly and published commercially in book form [Ref. 14]. This book made available to commercial industry and other interested persons background information on the DARPA program and on relevant studies in the Soviet Union.

C.   COMMERCIAL TECHNOLOGY TRANSFER ACTIVITIES

Interfaces were established with two commercial organizations that specialize in bringing together research laboratories and industrial cor-porations to exploit application of new technologies. SPC briefed the pres-idents of both Technology Transfer Conferences, Incorporated (TTCI) and TECHNOMIC Publishing, Inc. (TPI) on the DARPA programs in SHS and DCT. Com-munication links were then established between these organizations and laboratory participants in the DARPA program.

---

[1]U.S. Army Materials Technology Laboratory.

TABLE 4

INDUSTRIAL COMPANIES CONTACTED REGARDING SHS TECHNOLOGY
AND POINTS OF CONTACT

Aerojet Ordnance Company
9236 E. Hall Road
Downey, CA 90241
Attn: Dr. Harry Pearlman,
  Materials Consultant

Aluminum Company of America
Research Center
ALCOA Center, PA 15069
Attn: Dr. Siba Ray, Ceramics
  and Refractories

AVCO Systems Division
201 Lowell Street
Wilmington, MA 01887
Attn: Dr. Thomas Vasilos, Room 3,
  107A

Babcock and Wilcox Company
Research and Development Division
P.O. Box Box 239
Lynchburg, VA 24505
Attn: Mr. Daniel R. Petrak,
  Research Supervisor

Bell Laboratories Inc.
600 Mountain Avenue
Murray Hill, NJ 07974
Attn: Mr. Man Y. Yan, Room 6C308

Cabot Corporation
Concord Road
Billerica, MA 01821
Attn: Mr. James Belmont

Callery Chemical Company
Division of Mine Safety Appliances
  Company
P.O. Box 429
Pittsburgh, PA 15230
Attn: Mr. James W. Popp

Carbide Inc.
Norlin Industries
Arona Road
North Huntingdon, PA 15642
Attn: Mr. S. Grosel, Plant
  Manager

Corning Glass
Sullivan Science Park
Painted Post, NY 14870
Attn: Mr. Rodney Frost,
  Manager, Ceramic Development; or
  Mr. Robert McNally, Manager,
  Ceramic Research

Ceradyne Incorporated
167810 Milliken Avenue
Irvine, CA 92714
Attn: Dr. John Negrych,
  Vice President, Technology

Chromalloy Glass Division
Chromalloy American Corporation
Orangeburg, NY 10962
Attn: Mr. Robert Kessler

Crucible Inc.
Compaction Metals Operation
RDI, McKee and Robb Hill Road
Oakdale, PA 15071
Attn: Mr. William B. Eisen,
  Vice President/General Manager

Degussa Corporation
Route 46 at Hollister Road
Teterboro, NJ 07608
Attn: Dr. M. Verbeek

Dow Chemical
Central Research,
  New England Laboratory
P.O. Box 400
Wayland, MA 01778
Attn: Dr. Iwao Kohatsa

TABLE 4 (continued)

Dow Chemical Company
Chemical Research
Midland, MI 48640
Attn: Mr. George J. Quaderer or
  Mr. Ray Roach

Dresser Industries
P.O. Box 19566
Irvine, CA 92713
Attn: Dr. Gerald Miller

Exxon Research and Engineering
  Company
Corporate Research
P.O. Box 45
Linden, NJ 07036
Attn: Dr. Benard Kear

Ferro Corporation
1 Erieview Plaza
Cleveland, OH 44131
Attn: Mr. D. F. Beal, Commercial
  Development Representative

Fiber Materials, Inc.
Biddeford, ME 04005
Attn: Mr. Roger Pepper, Director,
  Advanced Materials Laboratory

General Atomic Company
P.O. Box 81608
San Diego, CA 92138
Attn: Dr. J. F. Watson, Director,
  Materials and Chemistry Division

General Electric Company
Re-Entry and Environment Systems
  Division
3198 Chestnut Street
Philadelphia, PA 19101
Attn: Dr. Peter Zavitsanos

Greenleaf Corporation
Greenleaf Drive
Saegertown, PA 16433
Attn: Mr. D. Keith Boyd,
  Treasurer

GTE Corporation
40 Sylvan Road
Waltham, MA 02254
Attn: Dr. George C. Wei

Kennametal Inc.
Research Laboratory
Greensburg, PA
Attn: Mr. George Rowland,
  Vice President, Research

Martin Marietta Corporation
Corporate Research Laboratory
1450 South Rolling Road
Baltimore, MD 21227
Attn: Dr. A Westwood, Associate
  Director

Micron Metals
7186 West Gates Avenues
Salt Lake City, UT 84120
Attn: Mr. Griff Williams,
  President

3M Company
3M Center, Building No. 251-2C-02
St. Paul, MN 55144
Attn: Dr. Ernest Duwell

Norton Company
1 New Bond Street
Worchester, MA 01606
Attn: Dr. M. Torti, Senior
  Scientist

Pfaudler Company
P.O. Box 1600
Rochester, NY 14692
Attn: Mr. Robert Naum

Ryan Metal Powder Technology
33661 James Pompo Drive
Fraser, MI 48092
Attn: Dr. C. Leznar

TABLE 4 (continued)

Shieldalloy Corporation
West Boulevard
Newfield, NJ 08344
Attn: Mr. George Campbell,
  Technical Director

Sohio Engineered Materials Company
Ceramic Research Center
P.O. Box 832
Niagara Falls, NY 14302
Attn: Dr. Jonathan J. Kim

Teledyne Firth Sterling
No. 1 Teledyne Place
Interchange City
La Vergne, TN 37086
Attn: Mr. T. Penrice,
  Vice President, Technology

United Technologies Research Center
Silver Lane, Mail Stop 25
East Hartford, CT 06108
Attn: Dr. Earl Thompson,
  Manager of Materials Sciences

Westinghouse Electric Company
1310 Beulah Road
Pittsburgh, PA 15235
Attn: Dr. D. E. Harrison,
  Manager, Materials
  Sciences Division

Dr. J. Birch Holt of LLNL participated in a TTCI conference that was held at Raleigh, North Carolina during February 1986. He discussed his work in SHS technology with representatives of U.S. industrial corporations.

When this report was in preparation, TPI, BTL, and LLNL were discussing a 2-day seminar on DCT to be held in Atlanta, Georgia. TPI anticipated that 50 to 70 industrial representatives would attend.

D. SHS INDUSTRIAL CONSORTIUM

LLNL and SPC have investigated the feasibility of establishing an industrial consortium to support the SHS program upon termination of DARPA funding at LLNL. A meeting was held on 26 March 1986 at LLNL with interested industrial representatives for discussions about formation of the consortium. A followup meeting will be held at LLNL in June 1986 with corporations that plan to participate in the consortium.

SPC contributed to this effort by establishing interfaces with commercial industry (see Section A) and by supplying background information to LLNL. Appendix B presents the briefing document on Soviet SHS technology that was prepared by SPC for this purpose. It contains copies of a series of vugraphs with relevant discussion. The presentation traces SHS history and discusses the Soviet development program and industrial applications. Also, details are provided on several types of processes, on processing equipment, and on Soviet patents that have been obtained in several countries.

III. Task 2: Technology Applications Within the Defense Community

A.   POTENTIAL APPLICATIONS IN PROCESSING ADVANCED CERAMICS

In reviewing the emergence of advanced technologies (Figure 1) and requirements for materials (Table 1), potential processing applications were identified for the three DARPA technologies as presented in Table 5. Emphasis among the items listed is on synthesis of advanced ceramics. Each of the three technologies also has the capability to combine synthesis of ceramic compounds with consolidation of dense monolithic products, which will be the ultimate payoff for development of these processing technologies.

1.   Advanced Ceramics

The review of emerging technologies was focused on potential applications of advanced ceramics in defense programs. Emphasis was on DCT since an interface group had been established previously between the SHS development program and defense systems programs, and the PCT development program was not considered to be ready for defense applications in the near term.

Advanced ceramics[1] usually are considered to be newer electronic or engineering ceramic products that have been developed since World War II. Traditional ceramics are usually considered to be older products that have been made primarily from mined minerals. Advanced ceramics is a fast-growing industrial commodity that is now classified according to uses by the Department of Commerce, as shown in Table 6. The 1984 U.S. market value for advanced ceramics was estimated at $4.7 billion [Ref. 17]. For the

---

[1]Sometimes also identified in technical literature as high-technology, high-performance, or fine ceramics.

19

TABLE 5

POTENTIAL USES FOR DARPA MATERIALS PROCESSING TECHNOLOGIES

Self-Propagating High-Temperature Reactions (to produce powders or mono-lithic shapes)
- Synthesis
  -- Binary compounds (borides, carbides, chalcogenides, hy-drides, intermetallics, nitrides, silicides, sulfides)
  -- Ceramic/ceramic microcomposites (e.g., as a duplex matrix or fibers in a homogeneous matrix)
  -- Cermets (e.g., cemented carbides)
  -- Gradated compositions
- Bonding
  -- Macrolaminates of dissimilar materials
  -- Surface cladding

Dynamic Compaction Technology (hot, cold)
- Consolidation of powders
  -- Compaction for densification and shaping (e.g., ceramics, cermets, intermetallics, metallics)
  -- Compaction for added strength (e.g, ceramic dispersants in metals)
- Phase transformation and synthesis
- Other uses
  -- Alteration of properties (e.g, electromagnetic, optical, physical)
  -- Porosity closure (e.g., in porous metal or cermet products)
  -- Surface protection (e.g., to provide wear resistance or oxidation resistance of metallic surfaces)

Plasmachemical Reactions (to produce powders and coat substrates)
- Synthesis
  -- Binary compounds (e.g, borides, carbides, intermetallics, nitrides)
  -- Complex ceramic compounds
- Surface modification
  -- For environmental protection
  -- For wear and abrasion resistance

TABLE 6

ADVANCED CERAMICS CLASSIFICATION BY END USE

Function/End Use	Four-Digit SIC Number	SIC Title
Heat Engines		
Vehicular engines	3714	Motor vehicle parts & accessories
Stationary engines	3511	Steam, gas, and hydraulic turbines
	3621	Motors and generators
Aircraft engines	3724	Aircraft engines and parts
Mechanical		
Cutting tools	3545	Cutting tools
Wear parts	3562	Ball and roller bearings
	3561	Pumps and pumping equipment
	3499	Fabricated metal products
Electronic		
Capacitors	3675	Electronic capacitors
IC packages/substrates	3674	Semiconductors
Resistors	3676	Electronic resistors
Other	3679	Electronic components, not elsewhere classified

Source: Reference 16

period 1984-85, the standard industrial classification (SIC) 3674 category of Table 6 is expected to have the highest annual growth rate (37.4 percent) among all 209 manufacturing industries listed by the U.S. Department of Commerce. By the year 1995, the Free World market level is predicted to be $17 billion [Ref. 18], and a level as high as $50 billion has been predicted for the year 2000 [Ref. 17]. Advanced ceramics of the 1990s may experience market growth similar to plastics of the 1960s.

A comprehensive illustration of the technology of advanced ceramics has been made available by the Fine Ceramics Office, Ministry of International Trade and Industry in Japan, as shown in Figure 2. This chart identifies performance functions and materials properties in the two middle circles, and 27 anticipated product applications are identified in the outer circle.

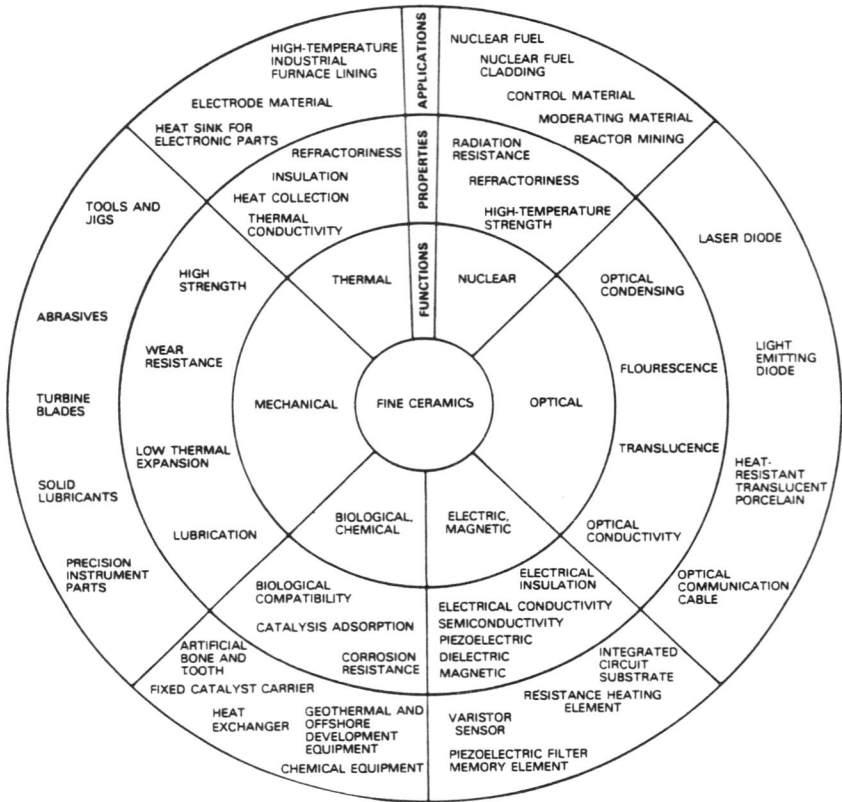

Source: Reference 19.

FIGURE 2

**FUNCTIONS, PROPERTIES, AND APPLICATIONS
OF ADVANCED CERAMIC TECHNOLOGY**

The processing techniques listed in Table 7 have been used most frequently to manufacture advanced ceramics. In addition, extrusion and
rolling should be included to make a more complete list. Many of the techniques listed, except for cold pressing and sintering, have not been employed extensively in processing traditional ceramics. Dynamic compaction
of advanced ceramic powders with explosives is even newer; therefore, it
has not been included in Table 7.

2. Dynamic Compaction of Ceramic Powders

Dynamic compaction is a processing technique with considerable market
potential for manufacturing a number of advanced ceramics. Key factors in
justification of this view are presented in Table 8.

Dynamic compaction can be performed either at room temperature or at
elevated temperatures that are considerably lower than those required for
hot pressing or conventional sintering of ceramic products. Lower temperatures promise an advantage in product purity since environmental contamination and chemical reactions among compaction components are eliminated as
major problem areas. The process is rapid compared to all alternative processes; therefore, the production time element becomes an advantage. Since
the explosive materials used for compaction have reproducible yields, product reproducibility is easily controlled. The constraint on large product
sizes normally associated with hot pressing is eliminated because only
space is required for processing; an expensive press bed is not needed.

The low capital investment requirement and capability for rapid manufacture of near-final shapes in a single operation indicate that dynamic
compaction should be cost competitive in situations where these factors can
be applied most effectively to gain advantage over alternative processes.
Perhaps the greatest competitive advantage can be realized in fabricating
either sizes that are too large or shapes that are not suitable for hot
compaction on presses. Conversely, overriding competition is anticipated in
situations where small pieces can be cold pressed at high rates and batch
sintered or where shapes are extremely complex, especially with sharp
changes in section thickness and surface contour (e.g., those often formed
by injecting molding).

TABLE 7

**TECHNIQUES USED TO MANUFACTURE
ADVANCED CERAMIC PRODUCTS**

- Cold Pressing--compacting the powders under extreme pressure at room temperature
- Hot Pressing--applying high temperature and great pressure simultaneously
- Sintering--heating powders under various atmospheric conditions and pressures down to a vacuum
- Reaction Bonding--binding the powders through a complex series of chemical reactions
- Hot Isostatic Pressing (HIP)--applying high temperature and high pressure simultaneously in three dimensions
- Injection Molding--molding the powders into the desired shape and then compacting through sintering or reaction-bonding
- Reaction Forming--interweaving fibers in a more solid matrix through chemical reactions (as in composites)

TABLE 8

WHAT DYNAMIC COMPACTION OF CERAMIC POWDERS OFFERS

Product Purity
- Little contamination from processing environments
- Little reaction among components in composites

Operational Aspects
- Rapid processing rate
- Excellent reproducibility
- Minimal limitations on size scale-up and aspect ratio

Cost Aspects
- Low capital investment
- Single process step to final shape

At the current stage of laboratory development, crackfree densified products have been compacted dynamically in the DARPA program. Product shapes have included flat platelets and other rectilinear forms, solid right circular cylinders, and tubes. Monolithic ceramic compounds that have been compacted include Al_2O_3, AlN, B_4C, sialons, SiC, TiB_2, and TiC. The compaction of microcomposites (e.g., Al_2O_3 in an AlN matrix) and a macro-composite (a steel mesh in AlN) also has been demonstrated.

B.   DYNAMIC COMPACTION OF ADVANCED CERAMICS AND COMPOSITES

This assessment demonstrates where technology transfer from laboratory parametric investigations into systems applications in U.S. Government pro-grams would be most propitious for DCT over the near term. The approach has been to discern among the many advanced ceramic products some specific items that are best suited for initial applications of DCT. When available, additional information has been included on preferred compositions for the products, on size and property requirements, on anticipated market value, etc.

The assessment has been subdivided into the following five applica-tional categories:

- Engines
- Armaments
- Heat exchangers
- Electronic hardware
- Mechanical equipment.

Selection of these categories has been influenced by predictions of high-volume production of advanced ceramics in the United States, Japan, and elsewhere. In addition, rapid solidification technology (RST) is included as another category. The combination of DCT and RST technologies shows promise for processing new classes of metastable materials that otherwise would not be achievable.

1.   Engines

The "ceramic engine" is included in every list of potential applica-tions for advanced ceramics. Ceramics promise higher engine operating tem-peratures—in some instances adiabatic operation of the engine—and higher operating efficiency; thus, the ultimate economic payoff is in fuel con-servation. All industrialized countries are developing ceramic components or ceramic-coated metal components for turbine or internal combustion engines. The development programs in Japan, West Germany; the U.S.S.R., and the United States probably are the most advanced.

The initial ceramic components, excluding the older spark plugs and a few other electrical devices, have been introduced only recently into production engines, and essentially all of them are oxide compositions. Typical applications in Toyota automobiles are listed in Table 9. Carbide and nitride compositions also have been used in prototypes of new engines. In those cases, the compositions generally have been silicon carbide, silicon nitride, and either silicates or zirconates.

Although property requirements for advanced ceramics will vary considerably with specific products, some generic requirements can be given as examples for many high-temperature structurals such as engine components. Those listed in Table 10 have been established as specific 10-year program objectives by the Engineering Research Association for High-Performance Ceramics in Japan. Emphasis is placed on strength and rigidity up to 1200°C, corrosion resistance in air up to 1300°C, and wear resistance up to 800°C.

In other specifications, minimum toughness properties at room temperature often are included among mechanical requirements. Values of K_{IC} on the order of 5 ksi $\sqrt{in.}$ or greater are typical of such requirements. For the U.S. Department of Energy's (DOE) engine development programs, rupture moduli in the range of 60 to 80 ksi and tensile strengths of approximately one-half of those values are considered to be minimum requirements.

The timeframe anticipated in the DoE program for introduction of nonoxide ceramic components in production engines is shown in Table 11. As indicated in the listings for the three time periods, complexity of product shape increases with time.

The market value for these advanced ceramic engine components is expected to grow rapidly over this same period. By the year 2000, the U.S. shipments are predicted to be $840 million, and the Free World market is predicted to be approximately $1 billion.

TABLE 9

CERAMIC COMPONENTS CURRENTLY USED IN TOYOTA AUTOMOBILES

Functional Ceramics

    Oxygen sensor (ZrO_2)

    Knock sensor (PZT)[a]

    Backup sensor (PAT)[b]

    Electric buzzer (PZT)[a]

    Thermal sensor for water temperature (Fe_3O_4-$CoMn_2O_3$-NiO)

    Thermal sensor for exhaust gas (Al_2O_3-Cr_2O_3)

    Blower resistor ($BaTiO_3$)

    Fuel level switch (Al_2O_3-Cr_2O_3)

    Heater for intake gas ($BaTiO_3$)

    Condenser ($BaTiO_3$)

    Motor core (Fe_2O_3-Mn_2O_3)

    Insulator for spark plug (Al_2O_3)

    Plate of hybrid integrated circuit (Al_2O_3)

    Plate of auto choke heater (Al_2O_3)

    Light-emitting diode (Ga-P)

    Electroluminescence (ZnS)

Structural Ceramics

    Mechanical seal of water pump (Al_2O_3)

    Catalyst pelleted support (Al_2O_3)

    Catalyst monolithic substrate (MgO-Al_2O_3-SiO_2)

    Ceramic fiber for fiber-reinforced metal piston (Al_2O_3-SiO_2)

    Heat insulator for catalyst (Al_2O_3-SiO_2)

---

[a]Piezoelectric zirconate titanate.

[b]Piezoelectric alumina titanate.

Source: Reference 19.

TABLE 10

PERFORMANCE OBJECTIVES FOR ADVANCED CERAMICS

Material Category	Performance Objective	Minimal Value

High-strength materials	$\geqslant$1200°C in air after 1000 hours holding:
	    Weibull modulus
	    Average tensile strength
		$M \geqslant 20$
		$\bar{\sigma} > 30 \text{ kg/mm}^2$

	$\geqslant$1200°C in air after 1000 hours continuous: Creep rupture strength	$\bar{\sigma} > 10 \text{ kg/mm}^2$

Corrosion-resistant materials	$\geqslant$1300°C in air after 1000 hours holding:
	    Weibull modulus
	    Corrosion resistance (weight gain)
	    Average tensile strength
		$M \geqslant 20$
		$\leqslant 1 \text{ mg/cm}^2$
		$\bar{\sigma} \geqslant 20 \text{ kg/mm}^2$

Wear-resistant materials	Room-temperature:
	    Wear resistance
	    Surface flatness
	800°C in air after 1000 hours holding:
	    Weibull modulus
	    Average tensile strength
		$\geqslant 10^{-4} \text{ mm}^3/\text{kg.mm}$
		$R \leqslant 2 \text{ }\mu\text{m}$
		
		$M \geqslant 22$
		$\bar{\sigma} > 50 \text{ kg/mm}^2$

Source: Reference 5.

TABLE 11

TIMEFRAME FOR CERAMIC APPLICATIONS IN ENGINES

Near Term (1980s)	Mid Term (late 1980s – early 1990s)	Long Term (post-1990)
Heat Exchangers 1000°C Clean environment	Heat Exchangers 1200°C Corrosive environment	Minimum Friction/Adiabatic Diesel Engines Pistons Bearings Combustion chamber
Wear Parts Seals Nozzles Bearings	Turbochargers Uncooled Diesel	Exhaust system Gas Turbine Engines Stators
Gas Igniters	Engine Components Piston liners, piston caps Heat plates, valve seats	Rotors Regenerators
Valves and Lifters	Precombustion chambers Exhaust path coatings	Aircraft Propulsion Engines
	Turbine Static Parts Combustors Shrouds	
	Turbine Rotors Auxiliary power units Rockets	

Source: Reference 2.

Examples of silicon carbide, silicon nitride, and aluminum lithium silicate components used in the development prototype of an automotive gas turbine engine that has been built for DOE are shown in Figures 3 and 4. Silicon carbide combustion chambers built for Volkswagenwerk AG are shown in Figure 5. A ceramic regenerator housing built for the Swedish Stirling engine is shown in Figure 6.

Components 4, 5, 6, and 9 in Figure 3, components 10 through 12 in Figure 4, and the components in Figures 5 and 6 are shapes that may be amenable to fabrication by dynamic compaction. The largest dimension of these components is a few inches, and section thicknesses are generally less than 1/2 inch. Similar ceramic shapes also have been considered for internal combustion engine components such as cylinder liners, piston caps, and valve heads.

Ceramic components from a roller bearing assembly are shown in Figure 7. This item is an example of a smaller engine component that is not suggested for application of DCT at this time. Alternative production processes are expected to be more cost competitive.

Although previous selections of ceramic compositions have been limited in engine applications, the number of different materials, as well as the different shapes and sizes, to be used can be expected to grow by the year 2000. As more information becomes available on properties of additional ceramic compounds (e.g., transition metal borides and carbides), nonoxide alternatives are likely to be selected more often for applications where a match is found between unique properties and requirements. In this regard, DCT should be widely adopted because it can be performed without much capital investment to investigate different types of materials.

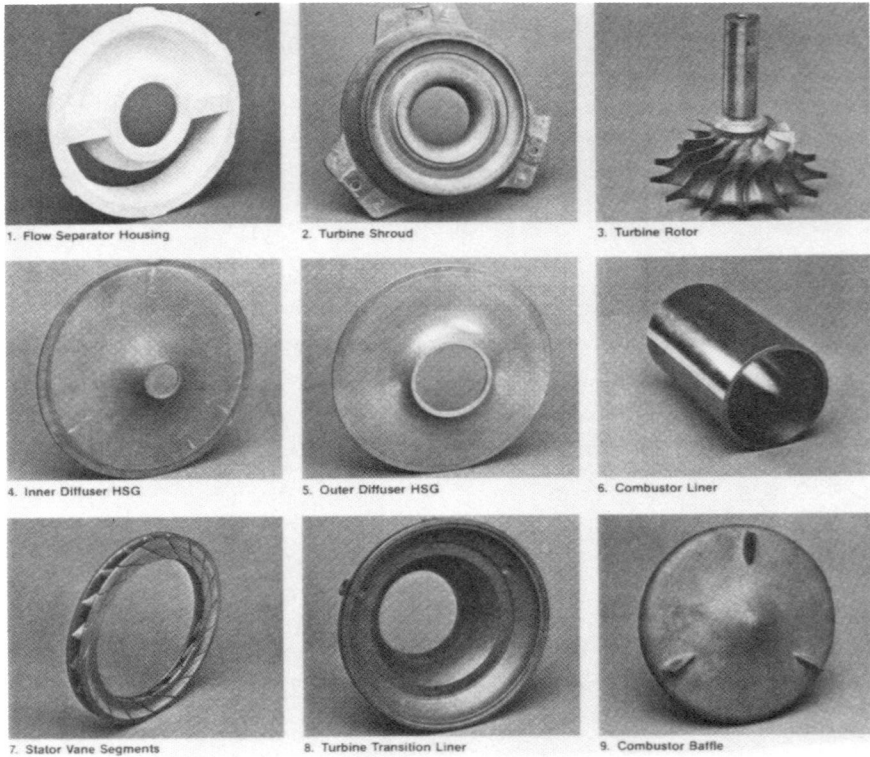

1. Flow Separator Housing   2. Turbine Shroud   3. Turbine Rotor

4. Inner Diffuser HSG   5. Outer Diffuser HSG   6. Combustor Liner

7. Stator Vane Segments   8. Turbine Transition Liner   9. Combustor Baffle

Note: See Figure 4 for identification of materials.

Source: Reference 20.

FIGURE 3

**CERAMIC COMPONENTS MADE FOR AN AUTOMOTIVE
GAS TURBINE ENGINE**

10. Turbine Backshroud

11. Bolts

12. Regenerator Shield

13. Alumina-Silica Insulation

Material identifications for components 1 through 12:

1.	LAS, RBSN	7.	RBSN, SiC
2.	SiC, RBSN, SSN	8.	RBSN, SiC
3.	SRBSN, SSN	9.	RBSN, SiC
4.	RBSN, SSN	10.	SiC, SSN
5.	RBSN, SSN	11.	RBSN, SiC, SSN
6.	RSSC, SiC	12.	RBSN, SiC

CERAMIC MATERIALS

SiC--Sintered Alpha Silicon Carbide   RBSN--Reaction-Bonded Silicon Nitride
LAS--Lithium Aluminum Silicate   SSN--Sintered Silicon Nitride
SRBSN--Sintered-Reaction-Bonded   RSSC--Reaction-Sintered Silicon
  Silicon Nitride   Carbide

Note: The material identifications are provided for components in
       both Figures 3 and 4.

Source: Reference 20.

FIGURE 4

**CERAMIC COMPONENTS AND MATERIAL IDENTIFICATIONS
IN AN AUTOMOTIVE GAS TURBINE ENGINE**

Source: Reference 21.

FIGURE 5

COMBUSTION CHAMBERS BUILT FOR A GAS TURBINE ENGINE

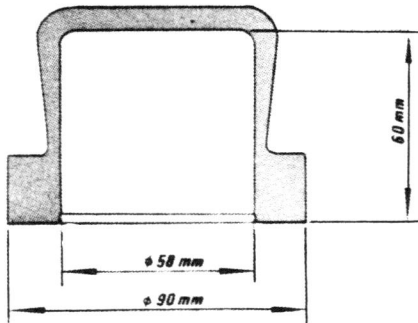

Source: Reference 22.

FIGURE 6

SILICON CARBIDE REGENERATOR HOUSING
BUILT FOR A STIRLING ENGINE

Source: Reference 23.

FIGURE 7

CERAMIC ROLLER BEARING ELEMENTS

2.   Armaments

Ceramic materials are mentioned most often in the armaments category for potential applications as lightweight armor. The ceramic products that are required for armor are usually particles or platelets, and in some limited applications cast products may be provided. Although many ceramic compositions have been investigated for these applications, details cannot be presented in unclassified literature. The armor shapes that would be most amenable to fabrication by the DCT process are platelets, cylinders, or simple variations of those configurations.

The production of composite armors, both micro and macro types, appears to be especially well suited for DCT. A major advantage in fabrication of macrocomposite armors is elimination of undesirable chemical reactions (between chemically incompatible materials) and assembly clearances. In some instances, cold bonding of components within a multilayer composite appears to be feasible.

A second potential application in the armaments category is production of kinetic energy (KE) penetrators with a composite structure. Emphasis in materials selections, especially for medium and large calibers, has shifted from very hard metals to fracture-tough, heavy metals in monolithic designs. However, the optimum penetrator may be one that combines the very hard, strong material with the heavier, tougher material, rather than one that compromises both of those requirements in a single material. DCT could be used, for example, to densify a ceramic compound as an insert or series of inserts inside a metal sleeve, as shown in Figure 8. The final product would have a relatively tough metal shell fitted around one or more slugs of a relatively hard, strong ceramic.

A third potential application for DCT in the armaments category is production of ceramic gun barrels or internal liners for metallic barrels. Some ceramic compounds (e.g., titanium diboride) are much more resistant to high-velocity gas erosion than metals. Since metal erosion is a major problem that limits the performance capability and service life of metal gun barrels, a full ceramic substitution or an inner ceramic surface for metal barrels is a rational solution to mitigate those problems. Dynamic compaction of an annular ceramic or cermet core inside inner and outer metal sleeves is envisioned. Both sleeves, or possibly just the inner one, would be removed in a post-compaction operation (e.g., by acid leaching).

Another military hardware item that has the tubular shape of a gun barrel but a greater length (approximately 40 to 50 feet) is the periscope tube used in submarines. Procurement of suitable metallic tubes has not been without problems [Ref. 24]. Application of DCT to development of a composite periscope tube would be a natural follow-on program to development of ceramic gun barrels or barrel liners. The compacted product could

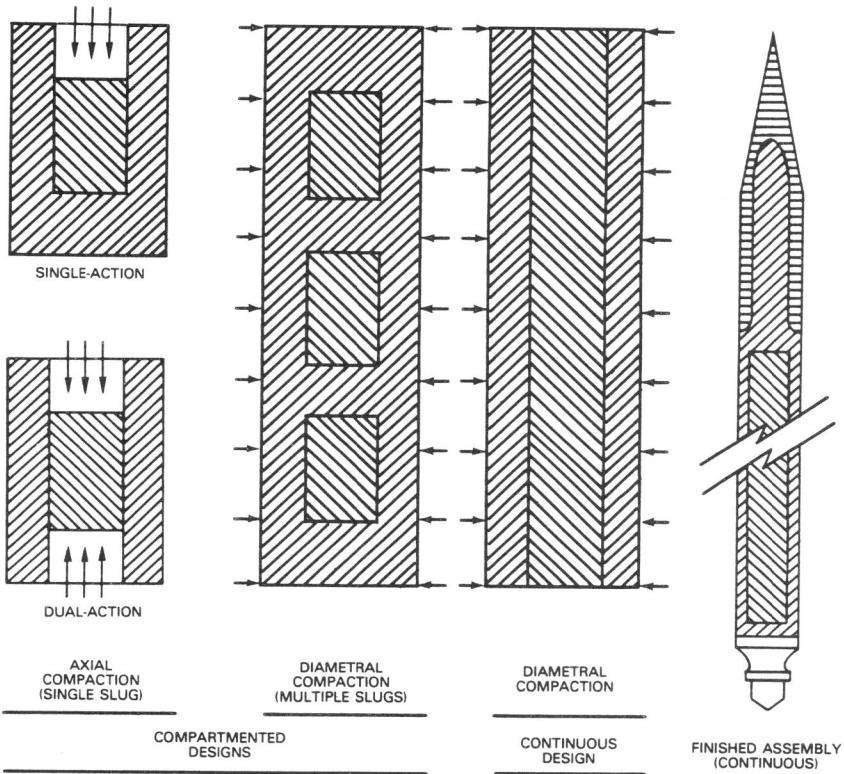

SINGLE-ACTION

DUAL-ACTION

AXIAL COMPACTION (SINGLE SLUG)	DIAMETRAL COMPACTION (MULTIPLE SLUGS)	DIAMETRAL COMPACTION	
COMPARTMENTED DESIGNS		CONTINUOUS DESIGN	FINISHED ASSEMBLY (CONTINUOUS)

POTENTIAL FABRICATION APPROACHES

NOT-TO-SCALE

METAL CASING   CERAMIC COMPONENT   WINDSHIELD

FIGURE 8

CONCEPTS FOR FABRICATION OF COMPOSITE
PENETRATORS BY DYNAMIC COMPACTION OF POWDERS

have inner and outer metal sheaths enclosing an annular ceramic core, as shown in Figure 9. Neither the tubular shape nor the length pose a problem for DCT. If the ceramic material for the design concept shown in Figure 9 were titanium diboride, the overall weight reduction in comparison to a stainless steel tube would be one-third (approximately 500 pounds). Furthermore, the dynamic compaction process is relatively inexpensive and requires relatively small capital investment compared to wrought processing by heavy forging, piercing, rolling, etc.

Once DCT applications such as composite gun barrels and periscope tubes have been mastered, another potential application to consider would be fabrication of ceramic or glassceramic (e.g., Pyroceram) cylindrical housings for deep-submergence structures. As depth of submergence is increased, advanced ceramics are expected to be preferred for fulfilling the severe requirements implicit in this application.

3.   Heat Exchangers

Heat exchanger designs of the type produced for a Brayton closed-cycle gas turbine engine, as shown in Figure 10, are considered to be too complex for fabrication by DCT at this time. On the other hand, simpler straight-tube designs are suitable for many industrial heat exchangers. In addition, designs required for heat pipe equipment are often straight tubes with a single wall thickness. Tube diameters are usually a few inches; wall thicknesses are on the order of 1/8" to 3/4"; lengths can vary from about a foot up to many feet. Service temperature may range from a few hundred to thousands of degrees. Service conditions for heat pipes used in cooling nuclear reactors with liquid metals involve operation in corrosive two-phase media. Both gaseous and liquid phases of potentially corrosive metals must be maintained under potentially erosive conditions at extremely high temperatures. Wicking components often are attached to the interior of such heat pipes, or grooves are required on the internal diametral surfaces.

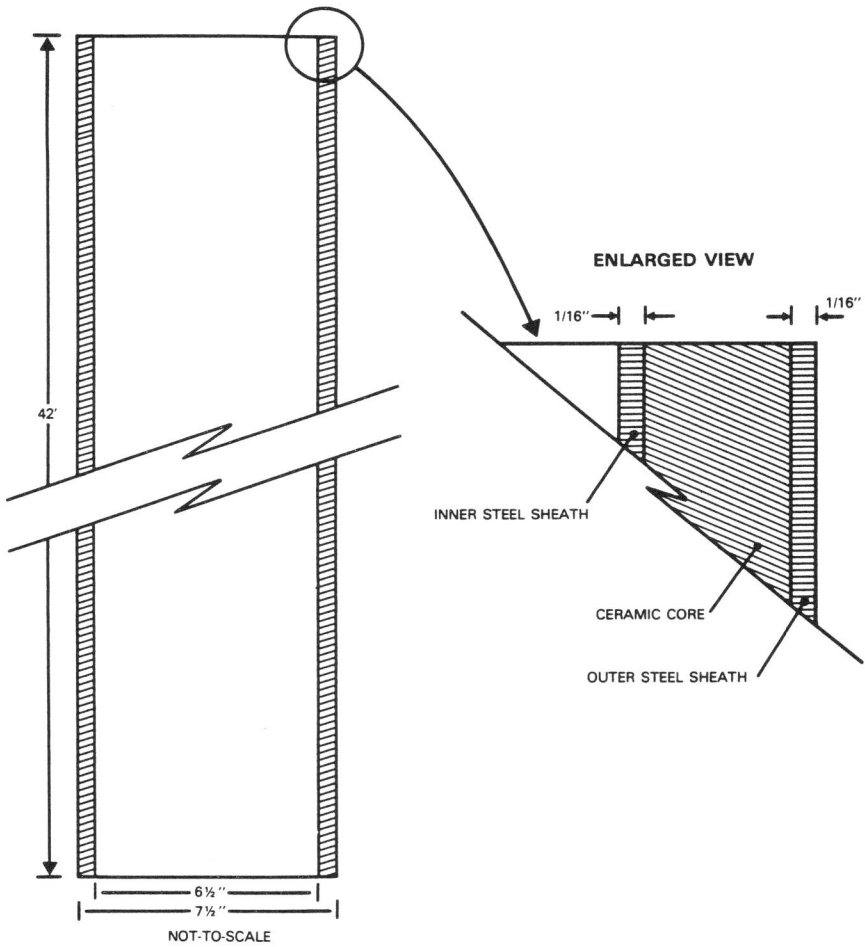

ENLARGED VIEW

1/16″

1/16″

INNER STEEL SHEATH

CERAMIC CORE

OUTER STEEL SHEATH

42′

6½″

7½″

NOT-TO-SCALE

FIGURE 9

CONCEPT FOR A COMPOSITE DYNAMICALLY COMPACTED PERISCOPE TUBE

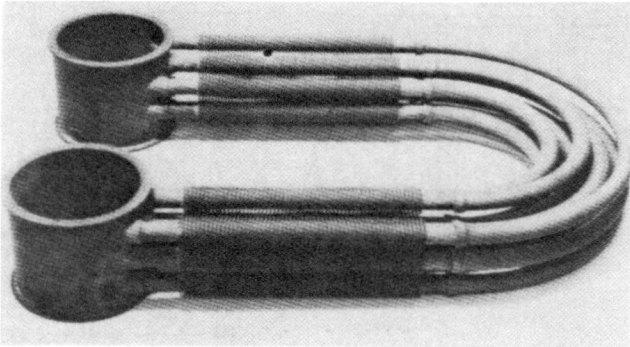

Source: Reference 11.

FIGURE 10

TEN-TUBE HEAT EXCHANGER ASSEMBLY FABRICATED
BY SLIP CASTING SILICON CARBIDE

Many different combinations of ceramic compounds or other materials in composites may be used in specific heat exchanger applications since the operating conditions vary widely. Compositions selected for several development prototypes for high-temperature service conditions have been silicon carbide and a lanthanum chromate cermet [Ref. 26].

The market value for ceramic heat exchangers is difficult to estimate because the potential applications are so diverse. Certainly, a significant portion of the market value predicted previously for ceramic components in heat engines is attributable to heat exchangers. In addition, the U.S. Department of Commerce has estimated that ceramic heat exchangers valued at $300 million are needed for recuperation operations in 12,100 industrial furnaces [Ref. 17]. The heat energy to be saved annually by those recuperation operations is estimated to be 100 trillion BTUs.

4.   **Electronic Hardware**

Electro-ceramic products include integrated-circuit packages and substrates, capacitors, sensors, transducers, resistors, piezoelectric components, magnetic components, and miscellaneous semiconductor devices. Among

these products, integrated circuits and capacitors make up by far the
largest shares of the expected total annual market value. For example,
shipments of electro-ceramics are predicted to be worth more than $3 bil-
lion in the United States by the year 2000, which is approximately 60 percent of the predicted value for all advanced ceramic products [Ref. 16].

The ceramic components in electronic hardware result in greater diver-
sity in performance and higher operating temperatures. Typical properties
that have expanded applications of electro-ceramics include tailored
dielectric constants, high piezoelectric coupling coefficients, high temperature stability, fast ion conduction, high magnetic permeability, trans-
parency to various signals, etc. To date, many of the ceramic compositions
in electronic hardware have been oxide compounds. As other compositional
types are more fully investigated, greater utilization of nonoxide compositions is expected. In some limited cases, electro-ceramic compounds have
been synthesized as well as shaped into products by DCT [Ref. 27].

The electronic product line of advanced ceramics appears to have only
limited potential in regard to initial DCT applications. Many of the products are quite small and are easily formed on automatic presses and densified by sintering. Many ceramic components are utilized as thin tapes, and
this shape is best formed by rolling. Logical applications for DCT include
magnetic components, a fairly heavy ceramic substrate for a microcircuit
(if an appropriate design could be found), or heavy components that serve
in the dual capacity of structural elements and resistors (see Figure 11).
Some magnets have been formed successfully by dynamic compaction of powders
[Ref. 29]. Also, magnets are now being formed from metallic glasses, as
discussed in Subsection 6 below.

5.   Mechanical Equipment

This functional category of applications is quite broad and to some
degree overlaps the other categories. Regardless, a few products are considered to be amenable to fabrication by DCT in the near term--e.g., machining and abrasion hardware, wear parts, seals, high-temperature structural components, etc. In some instances, ceramic compositions may be

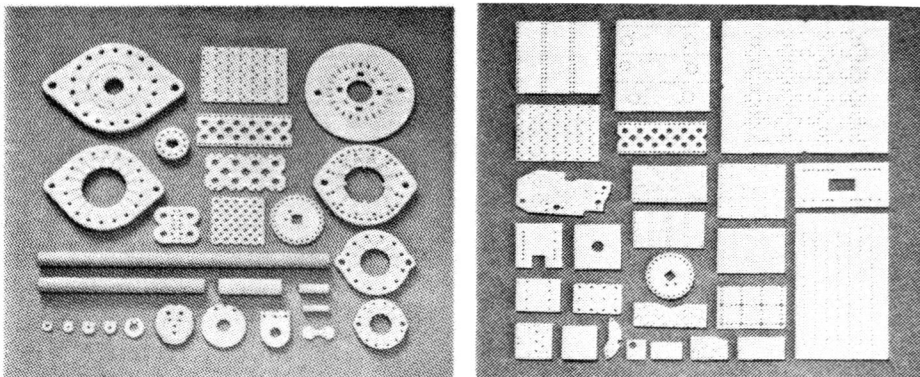

Resistor Components

Substrates for
Integrated Circuit Package

Source: Reference 28

FIGURE 11

EXAMPLES OF CERAMIC SUBSTRATES AND RESISTOR COMPONENTS

compacted with other materials in composite structures by DCT. In production of cubic boron nitride (CBN) products, the DCT process may be capable
of causing the transformation needed to make the required hard phase as
well as to form the desired product shape.

The broad utilization of ceramic materials in machining and abrasion
hardware is illustrated in Figure 12, which shows a variety of carbide and
nitride compositions. One of the compositions used in these applications in
the U.S.S.R. is titanium carbonitride (TiCN), which appears in the upper
left section of the figure.

In most mechanical applications, ceramics are used as monolithic
shapes or in composite structures. DCT may be used effectively in producing
composite structures--e.g., cemented carbide tools and bonded grinding
wheels. The ability to densify such products at ambient temperature should
contribute appreciably toward cost competitiveness in an industrial market.
In a few instances, the ceramic is used only as a coating--usually for surface protection against corrosion or wear.

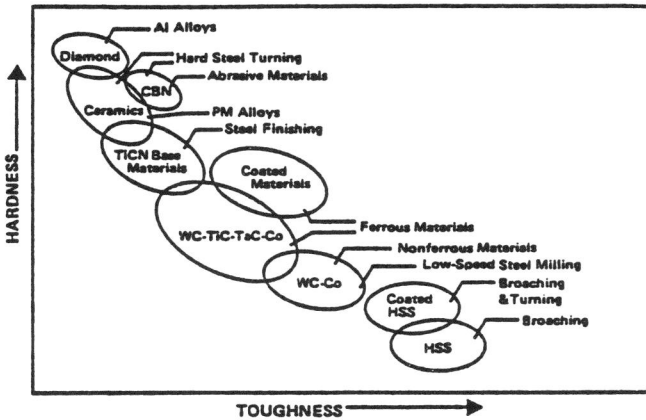

HSS = high-speed steels.

Source: Reference 30.

FIGURE 12

CERAMIC MATERIALS USED IN CUTTING AND ABRASIVE HARDWARE

Ceramic seals in pumps are another potential application for dynamic compaction. Such products are often simple in shape, as shown in Figure 13; larger sizes probably would be most suitable as the choice for fabrication of prototypes.

The use of structural components required for wear and erosion applications is an expanding field for ceramic products. Nozzles, tubes, and other hardware used in metal melting, refining, and casting operations are examples of high-temperature structural and erosion applications. Bearings (Figure 7) are an example of a structural and wear application.

The market share of mechanical hardware for advanced ceramic products is difficult to estimate because the products are used in so many different industries. Regardless, the market value is expected to be quite large by the year 2000, and the share will probably rank second only to electroceramics within the total market.

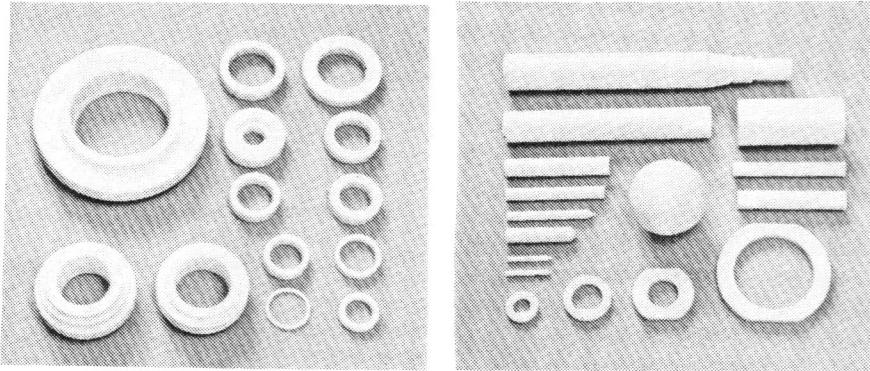

Seals   Pump Components

Source: Reference 28.

FIGURE 13

CERAMIC SEALS AND PUMP HARDWARE

6.   Rapid Solidification Technology

Rapid solidification is a relatively new technology by which materials that normally are expected to be crystalline are made amorphous directly from the molten state. These amorphous materials promise unique properties and metastable phases that previously have never been achieved.

The RST materials usually are solidified as powders or as ribbons, and the ribbons usually are converted to powders. The powders are then formed into useful product shapes. Wide utilization of RST in fabricating metal, ceramic, and composite products has been inhibited by the need for heating in conventional forming of products [Ref. 31]. Heating causes crystallization and grain growth, and the advantages attributed to amorphous or extremely fine grained materials are lost.

The combination of DCT with RST can circumvent the problem associated with heating in forming final products. Dynamic compaction is extremely fast and often can be performed at ambient or relatively low temperatures. The potential for forming and retaining metastable phases therefore is made evident.

The applications for RST materials are virtually infinite. Exploitation is limited only by investigations of new structures, phases, and properties. Two products that are appropriate for initial applications of DCT in this field are turbine blades and magnets. Superalloy and ceramic compositions, intermetallics, and composite structures all could be formed into the shapes of turbine blades. For example, some highly refractory intermetallic compositions ($Ni_{70}Ta_{30}$ and $Ni_{60}Nb_{40}$), which have been investigated just recently, look especially promising for turbine blade applications [Refs. 32 and 33]. Compositions that have been considered recently for applications in advanced magnetic devices include samarium-cobalt, neodymium-iron, and a variety of more complex combinations of elements. One such complex combination ($Fe_{40}Ni_{40}P_{14}B_6$) already has been dynamically compacted in the U.S.S.R. [Refs. 34 and 35].

One contractor, Teledyne-Wah Chang, is producing niobium alloy powders by RST for applications in the SP-100 and multimegawatt nuclear space power programs. SPC sent a briefing package on DCT to their sales manager to encourage consideration of this DARPA development for these applications.

C.   DEFENSE PROGRAM INTERFACES FOR DCT

Briefing information on the DCT program, which was prepared in cooperation with DARPA, is presented in Appendix C. This presentation identifies the laboratory participants in the program and describes accomplishments in compaction of ceramic and composite products and in computer modeling of dynamic processing. Pictures are shown of densely (more than 95 percent of theoretical limits) compacted products and their microstructures. The presentation was used in establishing contacts between the laboratory development program and defense systems programs.

The contacts with systems programs were initiated after an extensive review of R&D at Government laboratories that might require advanced products in the applicational categories identified in Section A. The Government offices and laboratories and the potential product applications that were compiled in this review are presented in Appendix D.

After a number of telephone contacts and briefings by mail, appropriate program contacts in DCT were identified, as listed in Table 12. Representatives from some of those programs attended DARPA's review of the DCT laboratory activities in September 1985. The laboratory participants and some of the other attendees identified in the table agreed to initiate planning for a follow-on systems program that will lead to a defense application of DCT.

D.   POTENTIAL SHS APPLICATIONS AND GOVERNMENT PROGRAM INTERFACES

1.   Domestic SHS Development Activities

Domestic SHS development activities have expanded considerably during the past 5 years, and participants now include a number of Government laboratories, universities, and industrial corporations. The two defense research agencies and seven Government laboratories that have been involved in the program are listed in Table 13. Although some of the participants have not performed major program tasks, the sheer numbers indicate that interest in SHS is widespread. As the number of participants increases, the potential for new ideas and new applications will multiply rapidly.

2.   Fabrication of Nickel/Aluminum Intermetallic Compounds

Aluminides are ordered intermetallic alloys that have great appeal in high-temperature structural applications because of their combined high strength and oxidation resistance [Ref. 36]. With recent development of increased ductility in these materials, specific applications are foreseen in ceramic engines and other programs.

Soviet research in fabrication of intermetallic compounds by SHS technology has extended over more than 20 years. Production applications have been cited for aluminides in the Soviet aerospace industry. Relevant background information on Soviet fabrication of aluminides by SHS technology has been prepared by SPC as briefing packages and submitted to the

TABLE 12

POINTS OF INTEREST IN DCT WITHIN GOVERNMENT PROGRAMS

Mr. P. D. Burke
Head, Technical Staff
Computer Sciences and
  Engineering Department
Naval Ocean Systems Center

Dr. James G. Early[a]
Acting Deputy Chief, Materials
National Bureau of Standards

Dr. Harry D. Fair
Tactical Technology Office
Defense Advanced Research
  Projects Agency

Dr. Thomas K. Glasgow
Lewis Research Center
National Aeronautics &
  Space Administration

Col. Robert Gomez
OSWR/OSD

Mr. James Humphrey[a]
OUSDR&E (TWP) OM
The Pentagon

Dr. R. N. Katz
U.S. Army Materials Technology
  Laboratory

Dr. W. Kitchens[a]
Terminal Ballistics Division
Ballistic Research Laboratory
Aberdeen Proving Ground

Dr. K. S. Mazdiyasni
Air Force Wright Aeronautical
  Laboratories

Dr. G. L. Moss[a]
Ballistic Research Laboratory
Aberdeen Proving Ground

Mr. Jerome Persh
OUSDR&E
The Pentagon

Dr. R. C. Pohanka
Code 431
Office of Naval Research

Mr. A. Schaffhauser
Manager, Conversion Technology
  Programs
Oak Ridge National Laboratory

Mr. Robert B. Schulz[a]
Department of Energy (CE)

Dr. Dennis J. Viechniki[a]
U.S. Army Materials Technology
  Laboratory

Mr. Andrus Viilu[a]
OUSDR&E (PWC/Land Warfare)
The Pentagon

---

[a]Attended review of DCT laboratory program in September 1985.

TABLE 13

**LOCATION OF DOMESTIC ACTIVITIES IN DEVELOPMENT OF SHS TECHNOLOGY
(Excluding Industrial Corporations)**

Facilities	Research Personnel
Defense Research Agencies	
● Materials Sciences Division, DARPA	S. G. Wax
● U.S. Army Research Office	A. Crowson
Government Laboratories	
● Lawrence Livermore National Laboratory	C. T. S. Chow
	C. Colmenares
	R. Gressman
	J. B. Holt
	D. Maiden
	C. McCaffrey
	G. Thomas
	D. Walmsley
● Lewis Research Center (NASA)	J. B. Hurst
● Los Alamos National Laboratory	R. G. Behrens
	M. A. King
	G. F. Metton
	S. M. Valone
● Naval Research Laboratory	D. Schrodt
	J. Wallace
	G. Y. Richardson
● Sandia National Laboratory (Livermore)	D. Hardesty
	S. B. Margolis
● U.S. Army Ballistic Research Laboratory	E. Horwath
	T. Kottke
	G. Moss
	A. Niiler
	M. Riley
● U.S. Army Materials Technology Laboratory	J. R. Alexander
	K. A. Gabriel
	S-S. Lin
	L. J. Lowder
	J. W. McCauley
	K. A. Moon
	H. Pevzner
	T. Resetar
	D. J. Viechnicki

TABLE 13 (continued)

Facilities	Research Personnel
Universities	
● Georgia Institute of Technology	K. V. Logan
	W. J. McLemore
	E. W. Price
● Massachusetts Institute of Technology	H. T. Brush
● Northwestern University	M. R. Booty
	B. J. Matkowsky
● Ohio State University	S. D. Dunmead
	C. E. Semler
● Rice University	L. Freidin
	G. P. Hansen
	J. L. Margrave
● State University of New York	J. Degreve
	V. Hlavacek
	J. Puszuski
● University of California-Davis	K. A. Philpot
	Z. A. Munir
● University of Illinois at Chicago	R. P. Burns
● Washington State University	S. Wojcicki

following program managers at Government agencies and laboratories and at one university:

- ● Dr. Robert G. Behrens, LANL (DOE)
- ● Dr. J. Birch Holt, LLNL (DOE)
- ● Dr. Harry A. Lipsitt, AFWAL, Wright Patterson AFB
- ● Dr. C. T. Liu, Oak Ridge National Laboratory
- ● Dr. Z. Munir, University of California, Davis
- ● Dr. Steven G. Wax, DARPA (MSD)
- ● Dr. Benjamin Wilcox, DARPA (MSD)

3. Ceramic Lining of Gun Barrels

The Soviets have used SHS to line metallic tubes internally with ceramic compounds [Ref. 37]. In this case, the ceramic compound is synthesized in a molten state and cast centrifugally inside a spinning metal tube. The reactants normally are a transition metal oxide, reducing agent, and nonmetallic elements, as shown by example below:

$$TiO_2 + C + 2 \, Mg \rightarrow TiC + 2 \, MgO$$

This procedure merits consideration in the program at the U.S. Army Materials Technology Laboratory to develop ceramic gun barrels or ceramic liners for metallic barrels. Potential benefits were discussed previously in Section B.2.

4.   Composite Kinetic Energy Penetrators

A metal/ceramic composite penetrator can possibly be fabricated by SHS technology. The potential benefits would be the same as those discussed previously in Section B.2, and the product would appear essentially the same as the concept illustrated in Figure 8.

A conceptual scheme for preparing a metal/ceramic segment is shown in Figure 14. The powder compact is reacted inside disposable graphite dies and a metal sleeve. Upon completion of the SHS reaction, the ceramic compound is densified by dual-action processing between molybdenum (or, perhaps, graphite) punches. The composite section is then machined to fit within a final assembly.

5.   Lightweight Ceramic Structural Tiles

Production of ceramic tiles for structural applications is another potential application for SHS. The concept is to densify the tiles by applying light pressure mechanically following the SHS reaction, as shown in a cross-sectional view in Figure 15. SHS technology would be much less expensive in fabricating tile structures than the standard hot pressing technique.

The powder load is prepressed to approximately 60 percent density. Ignition of the SHS reaction could be accomplished either by passing a current through wires placed in the grooves of the lower platen or by passing a low-voltage charge through the electrodes, as reported in a Soviet patent to produce tool bits [Ref. 38]. Once the combustion reaction is complete, light pressure is applied through the electrodes.

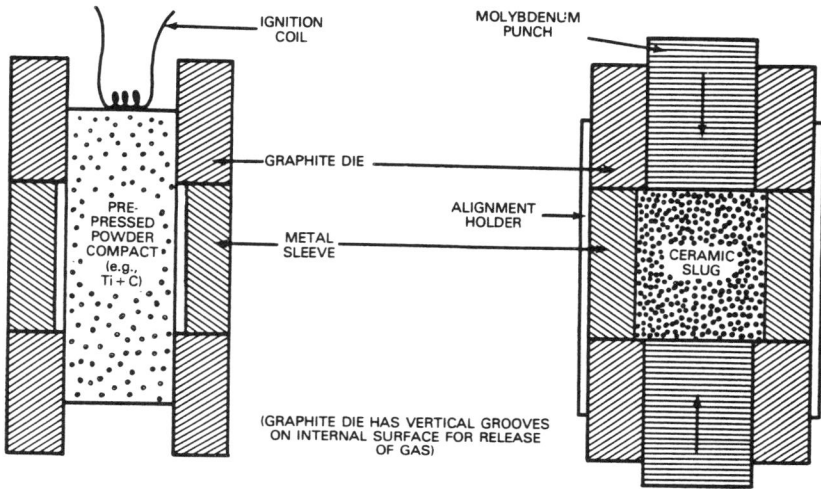

(1) PREREACTION ASSEMBLY

IGNITION COIL

MOLYBDENUM PUNCH

GRAPHITE DIE

PRE-PRESSED POWDER COMPACT (e.g., Ti + C)

ALIGNMENT HOLDER

METAL SLEEVE

CERAMIC SLUG

(GRAPHITE DIE HAS VERTICAL GROOVES ON INTERNAL SURFACE FOR RELEASE OF GAS)

(2) POST REACTION COMPRESSION

CROSS SECTIONAL VIEW, NOT-TO-SCALE

(3) MACHINED COMPOSITE SEGMENT

FIGURE 14

CONCEPTUAL SCHEME FOR PREPARING A COMPOSITE METAL/CERAMIC STRUCTURAL SEGMENT BY SHS

The potentially lightest structural tiles are beryllium boride com-positions; beryllium and boron have atomic numbers 4 and 5, respectively, in the periodic table of elements. As shown in Table 14, some beryllium boride compositions are characterized by high hardness and high melting points as well as low density; these properties are normally considered to be desirable for many structural applications. Development of beryllium boride has been prohibited in the Department of Defense because of poten-tial toxicity problems associated with beryllium. Unfortunately the tox-icity directive is misapplied in this case. Beryllium is not a toxic hazard when in massive form; in fact, it is used thusly without endangering the public in many commercial applications. During fabrication, standard pro-tective measures such as those used in handling all toxic materials (e.g., plutonium and mercury) would be in effect.

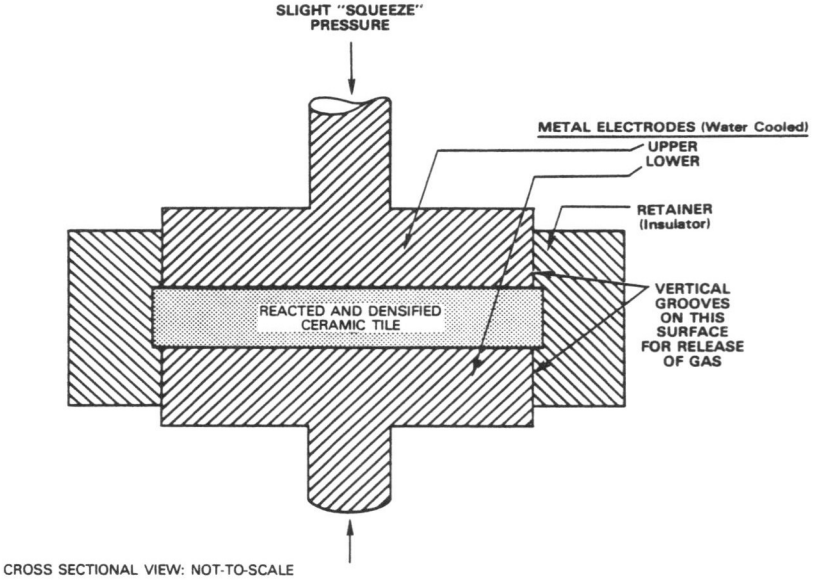

FIGURE 15

CONCEPTUAL SCHEME FOR DENSIFYING A CERAMIC
TILE BY SHS TECHNOLOGY

TABLE 14
SELECTED PROPERTIES OF TWO BERRYLLIUM BORIDE COMPOUNDS[a]

Compound	Density (g/cm^3)	Melting Point ($°$ C)	Hardness (HK_{100})
BeB_2	2.42	>1970	3180
BeB_6	2.35	2020-2120	2580

[a]Various sources of data.

Interfaces between the SHS program and defense programs have been established at the U.S. Army Materials Technology Laboratory (Drs. J. W. McCauley and D. J. Viechnicki) and Ballistic Research Laboratory (Mr. A. Niiler). Laboratory research is being sponsored by both organizations.

6.   Fabrication of Memory Alloys

Memory alloys are intermetallic compounds that have the capability of changing shape in response to changes in temperature. Memory alloy products have not been used extensively in the U.S. defense industry, but a number of aerospace applications are claimed in the U.S.S.R. In those applications, the memory alloy products are connectors on fuel lines or other service lines where junctures must be made in spaces that are extremely cramped for manipulating conventional mechnical devices (e.g., wrenches). Fabrication of titanium nickelides and copper aluminides by SHS technology is under investigation in the U.S.S.R. because several compositions in these systems are important memory alloys.

In conventional synthesis of titanium nickelides, complex facilities are required, power consumption is high, and product reproducibility is poor. In early experiments, V. I. Itin, at the Siberian Physicotechnical Institute, mixed nickel (PNK-OP4) and titanium powders, cold-pressed preforms to densities between 40 and 80 percent, and reacted them in a constant pressure bomb [Ref. 39]. Itin also added alloying elements (iron, cobalt, and aluminum) to control shape recovery temperature in the nickelide products. In follow-on studies, Itin concluded that manufacture of titanium nickelide and its alloys by SHS was more efficient than by the previous conventional method and that the process could be used to produce industrial shape memory alloys for which properties had been predetermined [Ref. 40].

In the latest study reported, Itin outlined a process for production of titanium nickelide by SHS [Ref. 41]. Dried powders of titanium and nickel were mixed, pressed into a blank, placed in a reactor, heated in argon to 300°C, and ignited [as previously reported in Ref. 39]. After synthesis, the blank was cooled and extruded at 950°C as a 25-mm-diameter rod. When the temperature dropped to 750°C, the rod was rolled into a strip of 1 mm thickness and annealed. The force generated during shape recovery of these alloys was greater than that found in materials produced by arc melting and plastic working. Itin was encouraged about using SHS technology to produce titanium nickelide but warned that the cost of starting powders must be reduced.

Itin and associates also have investigated SHS to produce memory alloys in the titanium-cobalt and copper-aluminum systems [Refs. 42, 43, and 44]. The SHS process was considered to be suitable for producing alumi-nide alloys, but not much encouragement was given for continuing investiga-tions in the other system. Some copper aluminide compositions are con-sidered to be suitable for use as electrodes in arc melting of metals as well as for memory alloy applications.

7.   Synthesis of Chalcogenides

Some chalcogenides are potentially important electronic materials (e.g., tellurides); others are being considered as high-temperature lubri-cants (e.g., selenides and sulfides). In the U.S.S.R., SHS has been under investigation as a production process for synthesis of chalcogenide com-pounds, and some applications of SHS chalcogenide products have been re-ported.

Since the current industrial methods used to prepare tungsten di-selenide as a lubricant are impractical for industrial products, A. G. Merzhanov and associates at the Institute of Chemical Physics, Soviet

Academy of Sciences, investigated SHS [Ref. 45]. The compound was synthesized in an argon atmosphere under pressures between 10 and 150 atmospheres. The crystal structure was investigated by X-ray spectral analysis, and the physicomechanical characteristics were determined after pressing without a binder. Results of the test showed that properties of the compound produced by SHS were equivalent to properties in conventionally produced (capsule synthesis) material.

V. K. Prokudina and associates, also of the Institute of Chemical Physics, studied the production of molybdenum disulfide by SHS [Ref. 46] because the compound has excellent lubricating properties. The material produced by conventional means is expensive and often contains unacceptable impurity levels of SiO_2 ·and residue from oil. The material that was produced under optimal SHS conditions has less impurities than industrial molybdenum disulfide. Prokudina concluded that SHS is suitable for high-output production and should be used as a basis for making industrial molybdenum disulfide.

8.   Combining Dissimilar Materials

Materials scientists who are knowledgeable about SHS technology forsee its ultimate potential to be in combining different materials in a single structure. The short SHS reaction periods and capability to localize intense heat lead to this prediction. The combination of materials could be brought about either by interface bonding on a macro scale or by forming composites on a micro scale. An example of a microcomposite that was synthesized and densified by SHS technology is the cemented carbide materials produced in the U.S.S.R. [Ref. 47].

Fabrication of a microcomposite in which SiC fibers are dispersed throughout an Si_3N_4 matrix merits investigation in the SHS development program. This material is considered to be extremely important as a refractory structural in domestic aerospace programs, and has been extremely difficult to prepare by reaction bonding and hot pressing [Ref. 48]. Alternative fabrication by SHS technology is feasible and is expected to be considerably simpler.

Nitride compounds have been synthesized by SHS technology at LLNL with solid sources of nitrogen (azides) [Refs. 49 and 50]. Since both Si_3N_4 and SiC sublime at high temperatures, SHS reaction under high pressure appears to be advisable. Therefore, a chemical reaction of the following form, under a pressure on the order of 50 atmospheres, would be reasonable:

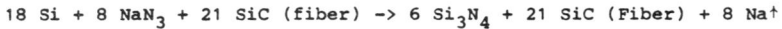

$$18 \ Si + 8 \ NaN_3 + 21 \ SiC \ (fiber) \rightarrow 6 \ Si_3N_4 + 21 \ SiC \ (Fiber) + 8 \ Na\uparrow$$

Another possiblity for obtaining a microcomposite with an Si_3N_4 matrix is suggested by the following reaction:

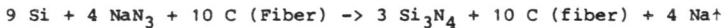

$$9 \ Si + 4 \ NaN_3 + 10 \ C \ (Fiber) \rightarrow 3 \ Si_3N_4 + 10 \ C \ (fiber) + 4 \ Na\uparrow$$

In this case, the fiber material would be graphite.

9.   Fabrication of Advanced Nuclear Fuels

The potential application of SHS technology for fabrication of advanced nuclear fuels has been brought to the attention of Government managers in fuels programs at Los Alamos National Laboratory (Dr. James Scott, SP-100 Program Office) and Westinghouse Hanford Company (Dr. Carl M. Cox), and a program interface has been established with Dr. R. G. Behrens at LANL. The reasoning behind this matchup of program needs and technology follows:

Program Needs:

- The advanced nuclear fuel programs need a nonoxide breeder fuel that is capable of high atom burnup at extremely high temperatures, has good thermal conductivity, and can be fabricated at reasonable cost.

- The advanced breeder fuels that have been subjected to irradiation testing have been fabricated by complex and expensive melting operations or treated in extremely high temperature and expensive furnance operations, with resultant lower process yields and higher costs than oxide fuel fabrication.

- When vacuum sintering is used, the fissile loading is reduced appreciably through volatilization.

- The present fabrication technology normally results in chemical contamination of the fuel composition unless painstaking and costly preventive measures are taken.

SHS Capabilities:

- Ceramic compounds (e.g., carbides and nitrides) are synthesized without a furnace (i.e., in a cold-wall vessel) and can be densified to 90-95 percent without a press.

- Any chemical contamination in presynthesis reactants is diminished during synthesis.

- Compositional adjustments (e.g., adding a transition metal such as Zr or combining C and N as a carbonitride) can be accomplished without significant process modification.

Compositional adjustment is listed above since Soviet research has revealed that the best nuclear fuels of carbide compositions may be those in which transition metal alloying elements are used to raise the UC melting point (e.g, 2500° to 3000°C) or nitride is added to form a carbonitride. In this regard, SHS reactions of the forms shown below are definitely feasible:

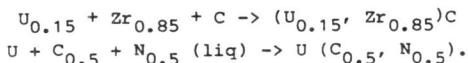

$$U_{0.15} + Zr_{0.85} + C \rightarrow (U_{0.15}, Zr_{0.85})C$$
$$U + C_{0.5} + N_{0.5} \text{ (liq)} \rightarrow U (C_{0.5}, N_{0.5}).$$

The simplest approach to demonstrate synthesis of a carbide nuclear fuel by SHS technology would be to produce a fuel powder. Large batches of carbide powders are produced routinely in the U.S.S.R. with essentially no process loss. For a preliminary demonstration of principle, a green compact of uranium, zirconium, and lampblack powders simply could be ignited in an inerted glovebox.

Carbide powder produced by SHS technology could be processed into pellets by the conventional processes that are used for fabrication of UO_2 fuels. However, dense fuel pellet shapes also can be obtained directly with SHS technology by simultaneous imposition of pressure with the combustion synthesis reaction. Soviet experience reveals that carbide product densities in the range of 90 to 95 percent can be obtained by restricting volume growth and controlling pressure of the inert gas medium during combustion.

In that case, the overriding advantage over current fuel fabrication technology would be elimination of both the melting (or furnace) operation and
the pressing operation in making pellets.

    Essentially full pellet density could be realized by applying pressure
following the SHS reaction. This concept is shown in Figure 16; it is an
experimental setup used at LLNL to densify a carbide compound other than
nuclear fuel.

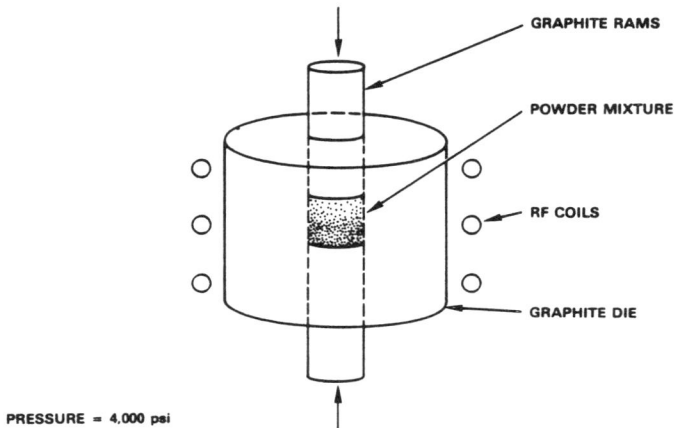

PRESSURE = 4,000 psi

FIGURE 16

PRESSING SETUP AT LAWRENCE LIVERMORE NATIONAL LABORATORY
FOR SIMULTANEOUS SOLID STATE REACTION AND DENSIFICATION
OF A CERAMIC COMPOUND

    Synthesis of advanced nitride fuels by SHS technology developed at
LLNL also is feasible [Ref. 49 and 50]. Three potential synthesis reactions
utilizing a solid source of nitrogen are illustrated below:

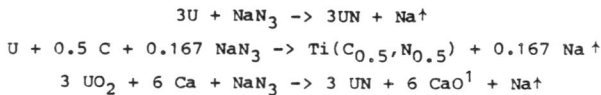

$$3U + NaN_3 \rightarrow 3UN + Na\uparrow$$
$$U + 0.5 C + 0.167 NaN_3 \rightarrow Ti(C_{0.5}, N_{0.5}) + 0.167 Na\uparrow$$
$$3 UO_2 + 6 Ca + NaN_3 \rightarrow 3 UN + 6 CaO^1 + Na\uparrow$$

---

[1]Subsequently, acid-leached from the nitride powder.

Similarly, an oxynitride composition or nitride/oxide composite fuel might be formed through the following reaction:

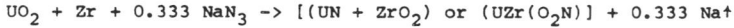

$$UO_2 + Zr + 0.333 \ NaN_3 \rightarrow [(UN + ZrO_2) \ or \ (UZr(O_2N)] + 0.333 \ Na\uparrow$$

Another concept for fabricating nuclear fuels by SHS technology is to produce long (continuous) fuel pieces rather than powder or the relatively short pellets, where length-to-diameter ratio is usually not more than 1.5:1. The long pieces would eliminate the problem with fuel ratchetting inside metallic cladding during service inside the nuclear reactor. Such fuel pieces possibly could be reacted and densified simultaneously on a drawbench, as shown in Figure 17, or perhaps in an extrusion press (without

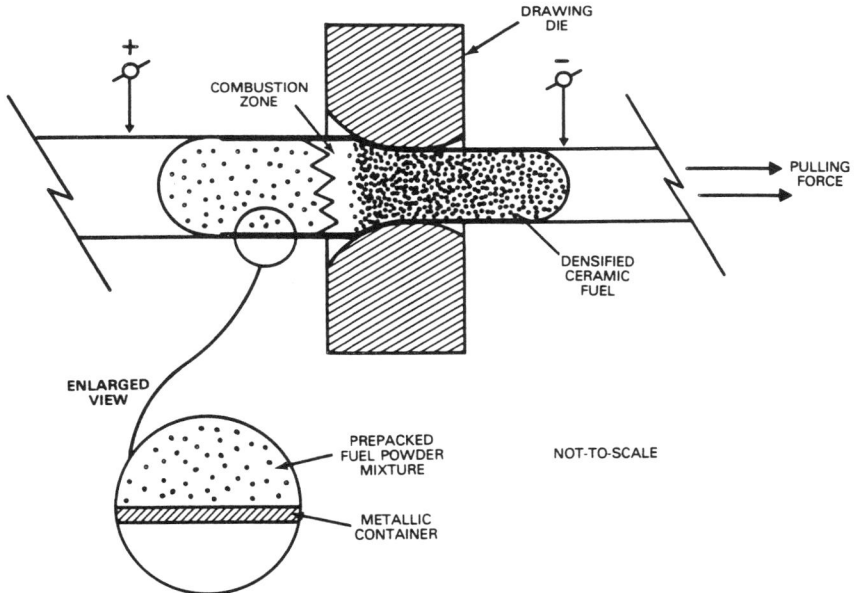

FIGURE 17

**CONCEPTUAL SCHEME FOR SIMULTANEOUS SYNTHESIS
AND DENSIFICATION OF LONG RODS OF CERAMIC NUCLEAR FUEL**

the metallic container around the fuel during processing). In this instance, ignition of the SHS reaction is accomplished with an electrical
current. The setup for ignition is quite similar to that used by Soviet
researchers in electroplastic deformation of metals on a drawbench [Ref.
51].

The metallic container around the powder mixture and reacted fuel
obviously will react to some degree with the fuel during combustion. The
alternatives are to remove it later by etching or to machine it and provide
a precision tolerance on the outside diameter of the fuel, and thereby have
the residual portion as a barrier to fuel/cladding reaction during service
in the nuclear reactor.

10. Isolation of High-Level Nuclear Wastes

The potential application of SHS technology to isolate nuclear wastes
in an inert matrix has been brought to the attention of Government managers
in waste programs in DOE (Office of Civilian Radioactive Waste Management,
Mr. William Purcell; Office of Defense Waste and Byproducts, Mr. David B.
LeClaire, Mr. W. A. Frankhauser, and Mr. Ray D. Walton, Jr.) and at Westinghouse Idaho Nuclear Company (Mr. J. R. Berreth). The reasoning behind
this matchup of program needs and technology follows:

Program Needs

- The oxide materials (glasses, minerals, and ceramics) favored
  currently in several national programs as the immobilization
  matrix are far from ideal selections in regard to their characteristic properties.

- Glasses are not high-temperature materials; consequently, cooling
  periods for spent fuel will need to be lengthened excessively to
  satisfy the severe temperature constraints that have been established for subsequent immobilization of high level wastes (HLW).

- Oxide ceramic materials (including the favored "synthetic" minerals) offer relatively little resistance to thermal shock.
  Again, constraints will be advisable in establishing thermal
  design requirements for HLW immobilization.

- Lattice structures in oxide compounds are not well suited to
  accommodate the many elemental transformations that result from
  radioactive decay during the lengthy immobilization period. Assurance of long-term chemical stability must be considered as a
  potentially difficult problem.

SHS Capabilities

- SHS technology is an operationally simple approach to synthesis of nonoxide refractory ceramic compounds.

- Nonoxide ceramic compounds (e.g., transition metal borides, carbides, and nitrides) have some unique properties that are more responsive than oxide properties to the HLW immobilization requirements.

Several transition metal borides, carbides, and nitrides are the most refractory among ceramic compounds. With proper selection of specific compositions within these groups, thermal stability in essentially all media (including air, water, acid, and molten metals) can be ensured up to relatively high temperatures. Also, many of these compounds are formed by metallic bonding and have desirable thermal characteristics similar to metals. Perhaps of greatest importance during HLW immobilization is the radioactive decay that forms different elements; the molecular defect structures of these nonoxide ceramics often will accommodate wide compositional variations in either metallic "host" or interstitial atoms.

Because of these characteristic properties, a compound such as titanium diboride (TiB_2) has considerable potential as a host matrix for HLW powders. It is one of the most refractory and chemically stable among transition metal borides. In addition, binary borides have been proven to be formed by essentially all metallic elements in the periodic table, and more complex compositions are formed with metals and other interstitial elements such as C, P, and O. With a gradated compositional mixture, as shown in Figure 18, the HLW powders might be chemically combined with TiB_2 in the central region of an immobilization mass as complex compositions. Also, the compositions could be changed gradually from that region outwards by adding decreasing amounts of HLW powders, with a protective region of pure TiB_2 provided on the exterior. This mass would have its maximum chemical stability on the exterior surface, and the thermal profile would be controllable.

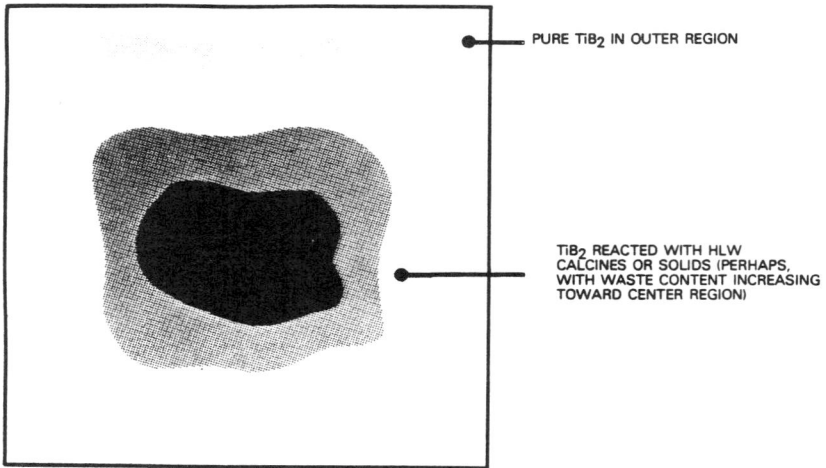

PURE TiB$_2$ IN OUTER REGION

TiB$_2$ REACTED WITH HLW
CALCINES OR SOLIDS (PERHAPS,
WITH WASTE CONTENT INCREASING
TOWARD CENTER REGION)

FIGURE 18

**CROSS SECTION THROUGH IMMOBILIZATION MASS
OF HIGH LEVEL WASTES (HLW)**

With SHS technology, nonoxide ceramic compounds can be synthesized without a furnace (as shown in Figure 19, for synthesis of titanium carbide powder) [Ref. 52]. After ignition of elemental reactants, extremely high combustion temperatures are exhibited; however, with rapid movement of the combustion wave (up to 15 cm/sec), the extremely high temperature region is restricted in volume. In Soviet SHS pilot production operations, powders of transition metal carbides and diborides have been synthesized in amounts of 100 tons per year in reaction vessels that hold up to 50 kg per processing batch. When the processing vessel is pressurized to prevent extraneous gas separation and growth in volume, the synthesis product is a compact of 90 to 95 percent density. When additional mechanical pressure is applied at a modest level immediately following passage of the combustion wave, product density greater than 99.8 percent is claimed.

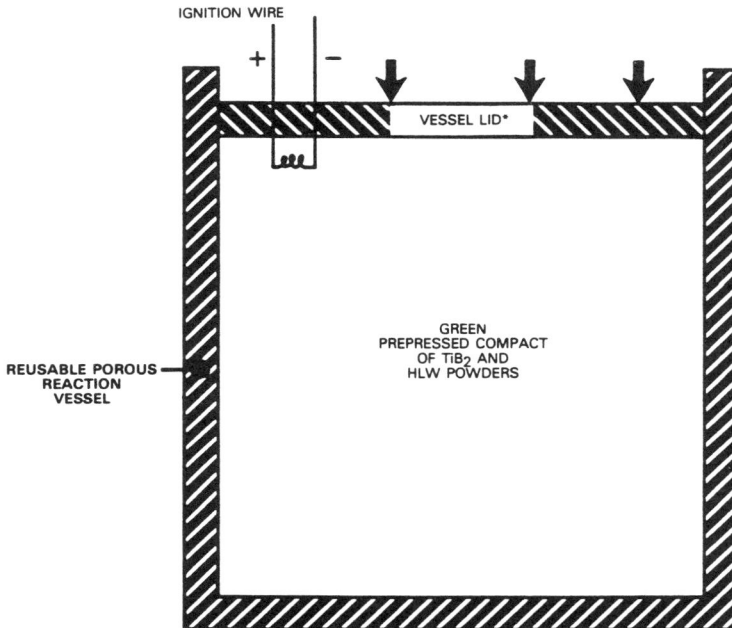

IGNITION WIRE

+   −

VESSEL LID*

REUSABLE POROUS
REACTION
VESSEL

GREEN
PREPRESSED COMPACT
OF TiB_2 AND
HLW POWDERS

* FLUID PRESSURE IS APPLIED ON VESSEL LID TO PREVENT GROWTH IN VOLUME,
AS AN AID IN DENSIFICATION DURING SYNTHESIS.

FIGURE 19

CROSS SECTION THROUGH REACTION VESSEL IN WHICH
TiB_2 WOULD BE SYNTHESIZED WITH A LOADING OF HIGH LEVEL WASTE

11.   Synthesis of High-Efficiency Radiation Shields

The search is on in several nuclear power programs for higher
efficiency radiation shields, where low mass and volume are major design
contraints. For example, in space applications, reactor shields for
neutrons must be more efficient per unit volume by several orders of
magnitude than shields normally placed around nuclear reactors at cental
power stations. Emphasis in the search for such shield materials has been
on transition metal and other hydrides with high hydrogen contents (e.g.,
zirconium, titanium, and lithium hydrides) that are laminated with other
materials in composite shield designs [Ref. 53 through 57].

In the U.S.S.R., transition metal hydrides have been formed by SHS technology in large tonnages for use in nuclear power programs. Although the technology has been described in numerous Soviet documents [Refs. 58 through 61], it has not been applied to synthesis of hydrides in the United States. Development is warranted in nuclear programs not only because of the capability to form hydride compositions but also because mixtures with multiple metallic constituents are readily synthesized, and hydrides could be laminated to other shield materials or to cladding elements during the synthesis reaction.

E.   POTENTIAL APPLICATIONS OF DARPA MATERIALS PROCESSING TECHNOLOGIES IN SPACE PROGRAMS

1.   Strategic Defense Initiative

Potential applications of DCT, SHS, and PCT in the SP-100 and multimegawatt space nuclear power reactors have been identified previously [Ref. 62]. The information has been distributed not only to program managers in the Office of the Strategic Defense Initiative (SDIO) but also to laboratory and contractor participants in the SDI space nuclear reactor programs. A few of the specific production applications that warrant attention of the SDI program are listed below:

- Advanced nuclear fuels (SHS technology)
- High-efficiency radiation shields (SHS technology)
- Coating refractory structurals to extend service temperature and life (PCT and SHS technology)
- Refractory heat pipes (DCT)
- Turbine and structural components (DCT, PCT, and SHS technology)
- Microelectronic and optical electronic components--e.g., in energy conversion and thermal disposal systems (DCT and SHS technology).

Briefing information on the DARPA materials processing technologies was provided, or presented orally, to the following persons who are involved in the SP-100 or multimegawatt power reactor programs or in materials and component development in support of those programs:

- Messrs. R. Verga and R. Wiley, SDIO, program management
- Dr. Ehsan Khan, DOE, materials development
- Mr. Elliot Kennel, AFWAL, Wright Patterson AFB, component and hardware development
- Lt. Dennis Kattner, Foreign Technology Division, U.S. Air Force, technology intelligence
- Drs. Thomas K. Glasgow and Stanley Levine, Lewis Research Center, NASA, materials development.

2.   Materials Processing in Space

Since the gasless combustion reaction had not been identified as an investigational field in materials processing in space, two potential experiments were suggested to the National Aeronautics and Space Administration (NASA) through contacts made at headquarters (Dr. Richard Halpern, Materials Processing Office) and at the George C. Marshall Space Flight Center (Dr. Robert Nauman, Materials Science Division). One experiment would assess the impact of zero gravity on gasless combustion reaction mechanisms; the other would investigate processing of a microcomposite material under the zero gravity condition.

The objective in the first experiment is to understand what happens to the combustion reaction when gravity-influenced convection is eliminated as a combustion mechanism. One possibility is that movement of the combustion wave would be slowed and process control would be facilitated.

The objective in the second experiment is to make a ceramic/ceramic microcomposite material in the molten state without the influence of gravity. The microstructure and properties of the casting are expected to be considerably different from material of the same composition that was cast in a gravity field. This experiment would be based on Soviet SHS technology wherein a reducing agent and oxide are used as reactants for melting and casting a ceramic compound [Ref. 63]. For example, a reaction to synthesize titanium carbide might be as follows:

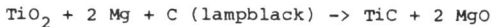

$$TiO_2 + 2 \ Mg + C \ (lampblack) \rightarrow TiC + 2 \ MgO$$

Normally, the carbide product (TiC) and oxide slag (MgO) are separated by centrifugal force. In the proposed experiment, the product and slag phases would remain intermixed as a ceramic/ceramic composite.

IV. Task 3: Relevant Soviet Materials Technology

Soviet literature on materials processing technologies was monitored routinely during the first seven quarters of the project work period and earlier in another DARPA project. This surveillance encompassed reviews of documents in both the Russian and English languages. In several instances, bibliographies were accumulated on specific subjects and recorded in prior SPC reports. Locations of the most significant bibliographies are recorded in Table 15 for future reference.

Soviet research in shock wave dynamics during the last five decades is considered to be first rate. Much of this research was a precursor to their more recent work in dynamic compression and synthesis of materials. Sources of Soviet literature that was published during the 1960s and 1970s on terminal ballistics, dynamic behavior of materials, fracture mechanics, and dynamic compression of materials are identified in the first section of Table 15. More recent papers on dynamic processing and modeling of material behavior also are identified in that section of the table.

SPC has developed an extensive file of Soviet SHS documents and papers from other countries. That file was published as a bibliography in 1984, as shown in the second section of Table 15, and additional Soviet papers were identified in the same report. An earlier report, Update of Soviet Technology in Production of Refractory Compounds (SPC 833, Working Paper), also contained a number of Soviet papers and patents on SHS technology.

Two bibliographies on the Soviet program in plasmachemical technology are referenced within the third section of Table 15. The first item contains information that is concerned primarily with R&D performed during the late 1970s and early 1980s. The second reference contains considerable information on industrial PCT applications in the U.S.S.R. during the 1980s.

65

TABLE 15

BIBLIOGRAPHIC INFORMATION ON SOVIET PROCESSING TECHNOLOGIES

Subject	Work Period of Report	Location in Prior SPC Project Reports
Dynamic Compaction Technology		
Terminal ballistics	Jun–Aug 1984	pp. C-10 to -12
Dynamic behavior of materials	Jun–Aug 1984	pp. C-12 & -13
Fracture mechanics	Jun–Aug 1984	pp. C-13 & -14
Dynamic compression of materials	Jun–Aug 1984	pp. C-15 to -17
Additional papers on dynamic processing and modeling	Jun–Nov 1984	pp. 138 to 140
Self-Propagating High-Temperature Synthesis Technology		
SPC file bibliography	Jun–Nov 1984	Appendix IV.4
Additional papers on SHS technology	Jun–Nov 1984	pp. 108-111A
Other SHS documents (SPC 833)	Sep 1982	--
Patents	--	Section I.A
Combustion theory	--	Section I.B
Processing technology	--	Section I.C
Contemporary press releases	--	Section I.D
Plasmachemical Technology		
Plasmachemical R&D	Jun–Nov 1984	pp. 56 to 59
Plasmachemical applications	Dec 1984 - May 1985	pp. IV-22 & -23

B.   KEY SOVIET PARTICIPANTS AND LITERATURE SOURCES

1.   Research Facilities and Personnel

Based on a continuing survey and assessment of Soviet literature, a number of research facilities and personnel who participated in each technology field that is included in the DARPA program have been identified. Since this information will be useful in future monitoring of Soviet literature, summaries have been prepared.

The research facilities that have participated most effectively in development of each technology are listed in Table 16. The first section, materials dynamics, identifies facilities that stress theoretical aspects of the technology; the second and third sections, SHS technology and PCT, identify facilities that, in general, worked both on theory and on development of potential technology applications.

The names of authors that have appeared most frequently in Soviet literature in regard to development of each technology are listed in Appendix E. The first list, for dynamics and shock waves, represents a broader field in dynamic energy applications than dynamic compaction of powders. Research in terminal ballistics, fracture mechanics, and dynamic materials processing is included. The other two lists, for SHS technology and PCT, mostly represent work performed in development of the specific process technology and associated equipment.

None of the information in either Table 16 or Appendix E is intended to be a complete listing of all facilities or names that were encountered in the literature survey. However, the information listed will serve as an excellent starting point for someone who wishes to conduct future literature searches in any of the fields represented.

2.   Technical Journals

More than 100 Soviet technical journals were examined to determine which were the better sources of relevant information. The 41 journals that proved to be most useful are listed in Appendix F. The 14 journals identified by an asterisk contained the most extensive information that was used in preparing prior SPC project reports. They would serve as an excellent starting point for future literature surveys and assessments.

C.   SPC TRANSLATIONS

More than 50 Soviet papers in the field of materials science have been translated by SPC and made available to participants in the DARPA advanced materials processing program.[1] In most instances, an SPC assessment of the

TABLE 16

SOVIET PARTICIPANTS IN
PROGRAMS RELEVANT TO DARPA TECHNOLOGIES

Materials Dynamics

Gidrodin Institute, Novosibirsk
Institute of Chemical Physics, Chernogolovka
Institute of Hydrodynamics, Novosibirsk
Institute of Materials Science, Ukrainian SSR
Institute of Problems of Mechanics, Moscow
A. F. Ioffe Physicotechnical Institute, Leningrad
Moscow State University

Self-Propagating High-Temperature Synthesis Technology

Baku Powder-Metallurgy Pilot Plant, Azerbaijan SSR
Branch Institute of Chemical Physics, Academy of
  Sciences of Ukrainian SSR
Institute of Chemical Physics, Chernogolovka
Institute of Materials Science, Kiev
Institute of Metallurgy and of Chemistry, Urals Science
  Center
V. D. Kuynetsov Siberian Physicotechnical Institute, Tomsk
  State University
Leningrad Technological Institute imeni Lensovet
Scientific-Research Institute of Applied Mathematics and
  Mechanics, Tomsk State University

Plasmachemical Technology

Alma-Ata Energy Institute
Institute for New Chemical Projects
Institute of Chemical Physics, Chernogolovka
Institute of Inorganic Chemistry, Latvian SSR
Institute of Materials Science, Ukrainian SSR
Institute of Metallurgy
Kharkov Higher Polytechnic Institute
Moscow Institute of Chemical Equipment Production

Soviet technology accompanied the translations. Titles of original Soviet papers and locations of the translations within prior SPC reports are recorded for future reference in Appendix G, which is organized as follows:

Section	Topic
1	Generic Materials Technology
2	Dynamic Compaction Technology
3	SHS Technology
4	Plasmachemical Technology

These SPC translations supplement information on Soviet processing technologies that is translated routinely by commerical sources from various technical periodicals. Many of the routinely translated papers are identified within the additional bibliographic information that was provided in Section A.

D. NATIONAL SOVIET SHS PROGRAM

A comprehensive review of the national Soviet SHS program by the director, Dr. A. G. Merzhanov, was translated by SPC and published in an earlier report [Ref. 64]. This review encompasses previous R&D findings, current program status, and future industrial aspirations. During the early 1980s, in Dr. Merzhanov's view, Soviet R&D in SHS technology was changing in emphasis from production of powders to dense monolithic shapes. At the time of that transition, program sponsorship was taken over by the Bureau of Tank Production. Since the change in sponsorship, essentially no information has appeared in open Soviet literature on simultaneous synthesis and densification of materials by SHS technology.

On the other hand, Soviet SHS technology for production of powders was made available to foreign countries through their Licensintorg organization in 1984. Kiser Research, Inc. of Washington, DC, is knowledgeable about the licensing process and has obtained powders produced by SHS technology in the U.S.S.R.

Information on the Soviet SHS program that was released by Licensintorg is provided in Appendix H. Item 1 discusses production capacity at the pilot plant in Chernogolovka; it also reviews production operations in nine industrial plants, which have reported annual production of over 1,000 tons each of titanium carbide, molybdenum disilicide, and silicon nitride. Item 2 reviews the R&D program at Chernogolovka and identifies many compounds that have been synthesized. Item 3 reports on specific industrial applications of SHS technology--both contemporary and future.

Recent press releases and technical documents [Ref. 65] identify the following processing areas as either current or future applications of SHS technology:

- Production of dense monolithic shapes
- Coating of substrates with refractory materials
- Joining of dissimilar components
- Production of metal hydrides for energy storage
- Production of composite materials (e.g., products from local rocks and titanium diboride/steel).

E. SOVIET PCT AND DCT DEVELOPMENT

PCT development also has status as a national U.S.S.R. program. As demonstrated in previous reports, development emphasis has been on design of new equipment for synthesis of ceramic powders. In some instances, these powders are sprayed directly onto substrates to provide surface protection. The most recent Soviet development in coating is use of a laser to create the plasmas [Ref. 66]. The feasibility of carbidizing and nitriding metal surfaces by this technique has been demonstrated at the Institute of Metallurgy imeni Baykov.

The Soviet emphasis in DCT development has been on synthesis and phase transformation to form materials with unique properties that are not otherwise attainable. Simple compaction of powders to form monolithic shapes has not received much attention; however, a Soviet paper published in 1985 may indicate a change of direction in this regard [Ref. 67]. This paper describes successful explosive densification of a cermet composition TiC/TiNi. Since this material is being used in the Soviet tool industry, the implication is that tool bits and similar products will be fabricated by DCT.

F.   APPROACHES TO INCREASING SERVICE PERFORMANCE OF REFRACTORY MATERIALS

Exploratory assessments of recent Soviet literature were made to identify their approaches for increasing service temperature and improving overall performance of refractory materials in the following areas:

- Transition metal alloys
- Fracture-tough ceramic and cermet materials
- Intermetallic (ordered) compounds
- Surface coating as an alternative to matrix strengthening.

Significant findings from those four assessments are reported in the remainder of this section.

A Soviet handbook of refractory compounds [Ref. 68] identifies those listed in Table 17 as heat-resistant materials (presumably, to be used for structural applications). Apparently, the book excludes refractory metals and intermetallic compounds such as aluminides, nickelides, etc. The first author of that book also originated the Soviet investigations of electron delocalization to produce complex ceramic compositions that exhibit unique ductility characteristics. One such material is listed within the borides category. The list also includes many cermet materials.

In a Soviet review of "modern" refractory compounds, the discussion concentrated on "complex multicomponent and heterogeneous materials based on binary refractory compounds" [Ref. 69]. The stated objective was to "explore. . .new heterogeneous materials and single-phase ternary composi-tions. . .that have capabilities extending beyond the binary types of oxygen-free refractory compounds."

TABLE 17

HEAT-RESISTANT MATERIALS FOR STRUCTURAL APPLICATIONS

All Beryllides

Can be used up to 1700°C

Borides

TiB_2-CrB_2 [a complex ceramic composition][a]

ZrB_2; identified as "borolitov" when used with Mo or Cr as a binder

CrB_2, HfB_2, NbB_2, TaB_2, VB_2, and CrB

Mo_2B_5 and W_2B_5; used as crucibles

Carbides

TiC [used extensively in the Soviet machine tool and die industry to replace WC][a]

TiC + TaC + NbC, with cobalt binder (20%)

TiC + TiB_2, with CoSi binder

TiC + Cr_3C_2, with Ni/Co/Cr binder

TiC, with steel binder [remains under study today][a]

CrC_2, with Ni binder

B_4C, with steel binder

SiC; used in combination with excess C, Si + C, Co, or B

Nitrides

TiN + MgO; has especially high resistance to thermal shock

BN [under intensive study in U.S.S.R.; converted to "hard" form by explosive compaction][a]

Si_3N_4; mostly used in combination with SiC, B_4C, or SiC/Fe

Silicides

Cr_3Si, with Cr binder

$MoSi_2$ [mostly for heating elements or refractory coatings on metals][a]

Ti_5Si_3 + SiC and $TiSi_2$ + SiC

---

[a]Comments in brackets are those of the author, Wm. Frankhouser of SPC, and are based on information other than that provided in the source document.

Source: Reference 68.

Boron carbide is considered to have a special place in the U.S.S.R. among nonoxide ceramic compounds because of its pure covalent bonding and high degree of valence electron localization. Soviet scientists believe that its inherent brittleness can be overcome by delocalization of electrons through introduction of other elements into the atomic lattice. This is the basis for their intensive development program in complex ceramic compositions. Other favorite materials in this program are modifications of TiC and eutectics consisting of transition metal carbides and excess carbon.

Additional Soviet statements pertinent to this discussion follow:

Lamellar Composites. Composites consisting of alternate layers of metal and oxide compound have the following advantages over ceramic matrix composites containing inclusions of dispersed particles or fibers [Ref. 70]:

- Metal layers are effective barriers against crack propagation
- Better thermal fatigue resistance
- Better impact strength
- Greater fracture toughness.

Transition Metal Refractory Alloys. The following limits for retention of high temperature strength have been cited for various strengthening mechanisms [Ref. 71]:

- Deformation strengthening, <0.4 T_M (melting temperature)
- Solid solution hardening, <0.5-0.6 T_M
- Precipitation strengthening (from a supersaturated solid solution), <0.6 T_M
- Dispersion strengthening, <0.7-0.9 T_M.

Soviet research in developing niobium alloys now combines dispersion strengthening and solid solution hardening effectively in new complex compositions.

Surface Protection of Refractory Alloys. Niobium alloys cannot operate effectively above 600° to 800°C in oxidizing environments. Coatings have been developed to extend the effective service temperature to the range of 1200° to 1300°C [Ref. 72].

Based on these brief assessments, a conclusive statement cannot be made about a single Soviet preference for the optimum refractory material. However, the following conclusions have been elicited from information reviewed:

Refractory Metals and Alloys

- Complex compositions that provide combined dispersion strengthening and solid solution hardening have been developed (and tested).

- Surface coatings are being developed to extend the service temperatures even higher.

- Welding and brazing processes are well developed for joining niobium alloys.

Complex Ceramic Compositions and Cermets

- Soviet scientists are convinced that ceramic compounds can be developed with adequate fracture toughness for most structural applications. Considerable progress is evident in modification of TiC and B_4C compounds with additive elements.

- The recent Soviet emphasis on cermet materials has been on TiC and Ti(C,N), with Ti/Ni and Ni/Mo binders.

Intermetallic Compounds

- Aluminides are considered to be important Soviet refractory materials. Emphasis has been on nickel and titanium aluminide base compositions, which are alloyed to reduce brittleness or to provide other desired properties.

Surface Protection of Refractory Alloys

- The preferred Soviet surface coatings are various combinations of aluminides and silicides that are alloyed with other elements.

- The preferred Soviet coating processes for the refractory alloys are plasmachemical spraying and diffusion impregnation from molten metal baths.

V. Task 4: Commercial Potential of Technology

A. BACKGROUND

1. Forecasting Base Cost and Price Values

New technologies are often ignored by program managers within the Government because cost is perceived to be too high. That perception usually is based on cost that is attributable to purchase of a few development components. Forecasting cost changes that may occur when proceeding from a development to a production phase is difficult; nevertheless, such predictions should be made available to potential users of advanced technologies.

The objective in this study is to develop production cost forecasts for the MSD technologies from information provided by laboratory participants in the program or found in the literature. These forecasts will demonstrate whether the specific advanced technology can compete with existing alternatives for processing materials in a production-scale environment. The approach is to develop a base cost (or starting point) and a base price and to forecast price behavior for a follow-on-production phase over the longer term. Forecasts of extended price behavior have been completed for production of a ceramic product in the form of cylindrical disks by DCT and flat tiles by SHS technology.

The base cost estimates were developed by calculating the annual cost of operating a facility that produces the ceramic product by the specific processing technology. Each facility was assumed to be an autonomous profit center within a relatively large corporation. The base price estimates were established by adding a profit margin and tax burden to the cost estimates. The estimates have been expressed in dollars per pound of contained ceramic compound in the product in order to facilitate comparisons with values

75

ascribed to previous sales of similar products that were fabricated by existing technologies.

2.   Assumptions Regarding Products and Annual Production Level

The end use of each product has been assumed to be ceramic components in a ceramic/metal assembly. The basis for this assumption is a matter of convenience in establishing reasonable annual production quantities for structural applications. The ceramic compound, TiB_2, was selected because it has been hot compacted by DCT at BTL and synthesized by SHS technology both at LLNL and at an SHS subcontractor laboratory. The final structural assembly is assumed to weigh 824 kg. Each assembly would contain about 41 percent (by mass) of ceramic components, or 341 kg of TiB_2.

The annual TiB_2 production requirement is based on fabricating 385 assemblies per year--a modest number for a structural application. Thus, the annual production requirement of TiB_2, by mass, would be about 131 MTs ([341 x 385] ÷ 1000). Assuming a density of 96 to 98 percent for TiB_2, this mass relates to annual production of approximately 87,000 disks or 50,000 tiles.

The process yield assumed in the cost calculations is 92.5 percent. The nonyield material includes allotments of 5 percent for scrap and 2.5 percent for samples and archive materials. Worker productivity in operational areas is assumed to be in the range of 65 to 70 percent.

The facilities are assumed to operate over 500 shifts per year (50 weeks x 5 days x 2 shifts). Cleanup and makeup operations for unexpected shutdowns would be performed on the third daily shift or over weekends. Two weeks are allotted to annual plant shutdown.

B.   **BASE COST AND PRICE ESTIMATES FOR DYNAMIC COMPACTION OF CERAMIC DISKS**

1.   **Product Description**

The product shape--a disk--was selected as a simple geometric con-figuration that would be amenable to fabrication by DCT, but which also has been explosively hot compacted at BTL. Although the size of the disk is amenable to fabrication by DCT, its dimensions would pose considerable difficulty for fabrication by standard hot pressing on a hydraulic or mechanical press.

As shown in Figure 20, the disk comprises a ceramic core within a thin steel sheath. Interbonding of the sheath components is accomplished during explosive compaction of the entire core/sheath assembly.

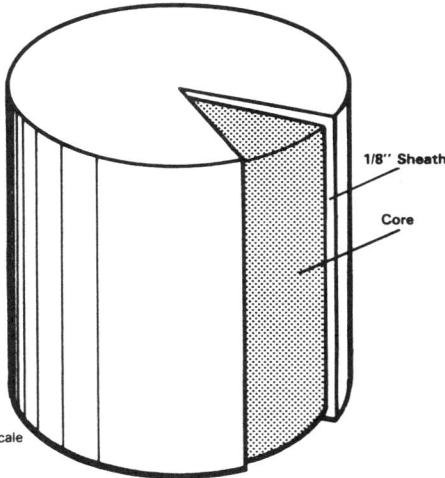

1/8" Sheath

Core

MATERIALS

• TiB_2 Core
• Low-Carbon Steel Sheath

DIMENSIONS (in)

	OD	HEIGHT
• Core	3	3
• Overall	3¼	3¼

WEIGHT OF TiB_2

• 1.51 kg (3.32 lb)
  (at 96% Density)

NOTE: Not-to-scale

FIGURE 20

EXPLOSIVELY HOT COMPACTED DISK
USED IN DEVELOPMENT OF COST ESTIMATE

If precise control of the external dimensions of the disk is required (e.g., when the product might be used as a component within a large assembly), only machining of the relatively soft outer sheath would be necessary. Machining or grinding of the outer surface of the extremely hard ceramic core would add significantly to cost of the final product.

Product density of approximately 93 to 98 percent would be obtained by hot compaction in the range of 600° to 1100°C. Should the highest levels of the product density range have to be combined with the lowest levels of the compaction temperature range, investigation of sintering aids would be appropriate to assess the potential for lowering temperature.

2.   DCT Process Basis for Cost Estimate

The basic assumption is that five disks are explosively hot compacted in a single firing assembly. The operational sequence for a single firing includes the following steps (which are illustrated with additional details in the flow diagram in Figure 21):

- Construction of the steel work container (which becomes the steel sheath around the diametral surfaces of the disks, after firing)
- Preparation of the green TiB_2 compacts
- Loading, evacuation, sealing, and heating of the complete compaction assembly
- Preparation and loading of the drop stand and firing assembly
- Explosive compaction (after transfer of the compaction assembly from furnace to drop stand)
- Disassembly and finishing operations.

The number of firings required per shift is 37 (87,000 disks/yr ÷ 0.925 ÷ 5 disks/firing ÷ 500 shifts).

3.   Process Materials Basis for Cost Estimate

Process materials have been grouped as TiB_2 powder, working explosive, and miscellaneous (steel and expendables). Specific materials items in these last two categories are identified in Figure 22.

		ASSEMBLIES AND COMPACTION LOADING		
		WORK CONTAINER	COMPACTION ASSEMBLY	DROP STAND AND FIRING ASSEMBLY	EXPLOSIVE LOADING

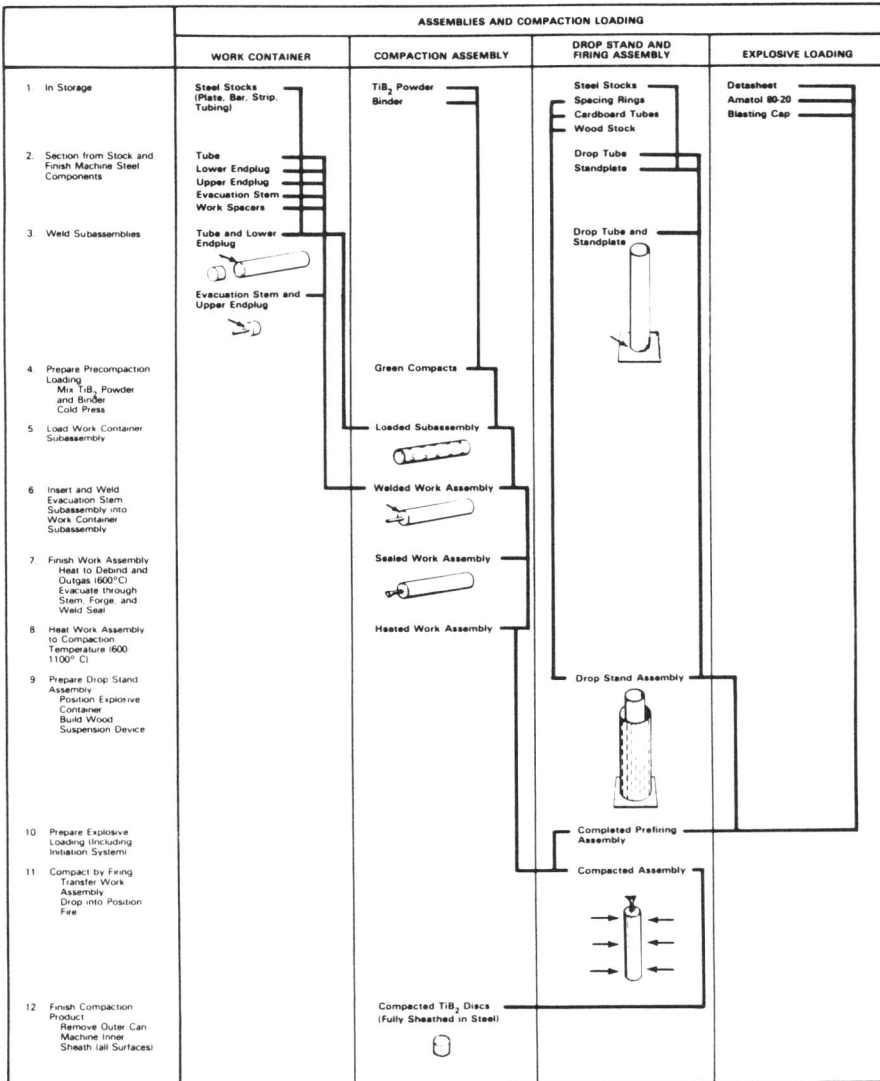

		WORK CONTAINER	COMPACTION ASSEMBLY	DROP STAND AND FIRING ASSEMBLY	EXPLOSIVE LOADING
1.	In Storage	Steel Stocks (Plate, Bar, Strip, Tubing)	TiB_2 Powder Binder	Steel Stocks / Spacing Rings / Cardboard Tubes / Wood Stock	Detasheet / Amatol 80-20 / Blasting Cap
2.	Section from Stock and Finish Machine Steel Components	Tube / Lower Endplug / Upper Endplug / Evacuation Stem / Work Spacers		Drop Tube / Standplate	
3.	Weld Subassemblies	Tube and Lower Endplug / Evacuation Stem and Upper Endplug		Drop Tube and Standplate	
4.	Prepare Precompaction Loading — Mix TiB_2 Powder and Binder, Cold Press		Green Compacts		
5.	Load Work Container Subassembly		Loaded Subassembly		
6.	Insert and Weld Evacuation Stem Subassembly into Work Container Subassembly		Welded Work Assembly		
7.	Finish Work Assembly — Heat to Debind and Outgas (600°C), Evacuate through Stem, Forge, and Weld Seal		Sealed Work Assembly		
8.	Heat Work Assembly to Compaction Temperature (600–1100° C)		Heated Work Assembly		
9.	Prepare Drop Stand Assembly — Position Explosive Container, Build Wood Suspension Device			Drop Stand Assembly	
10.	Prepare Explosive Loading (Including Initiation System)			Completed Prefiring Assembly	
11.	Compact by Firing — Transfer Work Assembly, Drop into Position, Fire			Compacted Assembly	
12.	Finish Compaction Product — Remove Outer Can, Machine Inner Sheath (all Surfaces)		Compacted TiB_2 Discs (Fully Sheathed in Steel)		

FIGURE 21

PROCESSING OPERATIONS FOR HOT COMPACTION OF TiB_2 DISKS

**WORK CONTAINER COMPONENTS
(LOW CARBON STEEL)**

Tube (1) (4-1/4"OD × 4"ID × 20")

Work Spaces (4)
(4" Dia × 1/4")

Lower Endplug (1)
(4" Dia × 2")

Evacuation Stem (1)
(3/8"OD × 1/8"ID × 12")

Upper Endplug (1)
(4" Dia × 2"; with Drilled
Hole for Evacuation)

DROP STAND ASSEMBLY COMPONENTS

Standplate (1)
(12" × 12" × 1/4";
Low Carbon Steel)

Drop Tube (1)
(4-3/4"OD × 4-1/4"ID × 30";
Low Carbon Steel)

Pine Boards (To Make Stand for
Suspending Work Assembly
Over Drop Tube)

EXPLOSIVE LOADING AND ACCESSORIES

Cardboard Tube (Chargeholder)

Tube (1)
(9-3/4"OD × 9-5/8"ID × 24")

Masonite Spacing Ring (1)
(9-5/8"OD × 4-3/4" ID × 1/2")

Explosive Initiation System
- Detasheet (1/2 lb)
- Blasting Cap (1)

Working Explosive: Amatol 80-20 (Powder) (30 lb)

NOTE: Not-to-scale

FIGURE 22

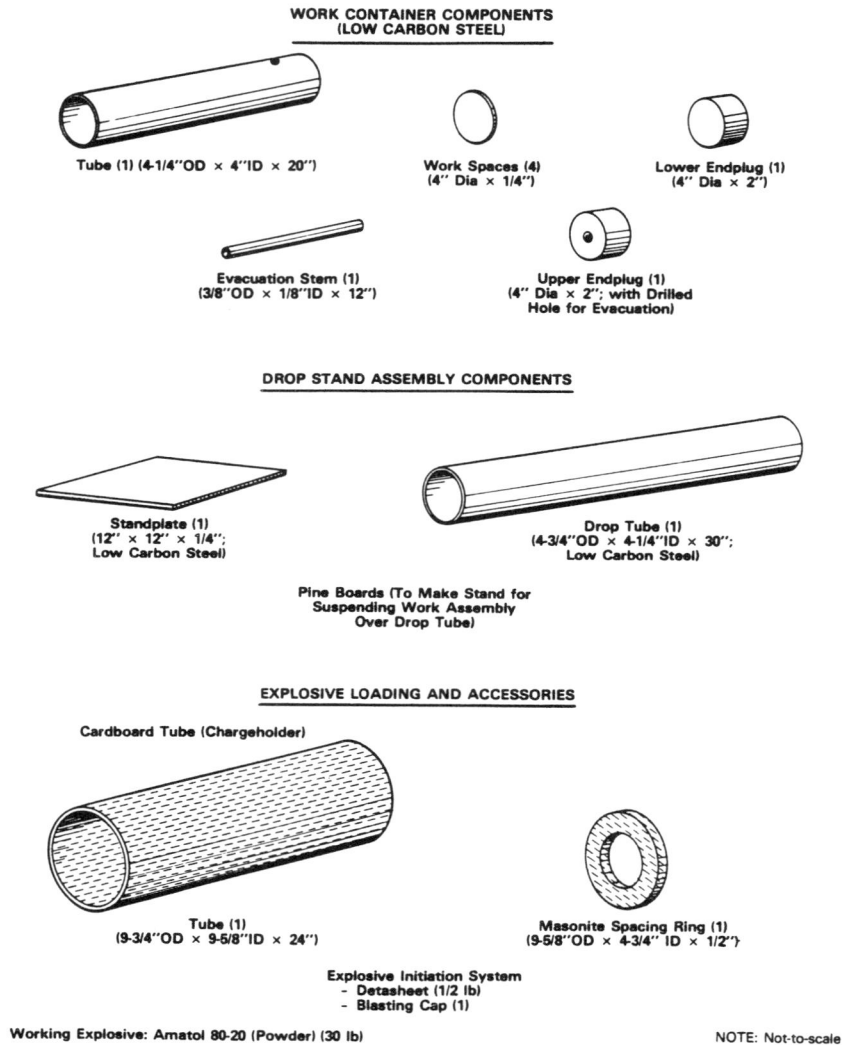

COMPONENTS FOR STRUCTURAL ASSEMBLY, EXPLOSIVE CHARGE,
AND INITIATION SYSTEM (FOR SINGLE FIRING)

Cost of materials purchases is developed in Table 18 (on the basis of a single explosive firing). The total materials cost is $392.20, of which 69 percent is attributable to TiB_2 powder and 23 percent to the working explosive.

TABLE 18

MATERIALS COSTS
(For Single Firing of Five Disks)

1. TiB_2 Powder

    $15 per lb x $3.32/lb x 5 disks = $249

2. Working Explosive (Amatol 80-20)

    30 lb (per firing) x $2.75/lb = $82.50

3. Miscellaneous

    1. Low-carbon steel [34 lb (per firing) x $0.50/lb] ÷ 0.7 (preparation yield) = $24.30

    2. Ignition system (including 1/2 lb Detasheet, blasting cap, cardboard tube, Masonite spacing ring, and wood) = $7

4. Total Cost per Firing

    [$249 + $82.50 + $24.30 + $7] ÷ 0.925 (process yield) = $392.20

4. Facility Basis for Cost Estimate

The facility is assumed to be a free-standing building of "Butler-type" construction. It is located away from population centers in the middle of a 200-acre tract. Specialized equipment in the building includes containment for explosive processing (storage magazines and firing closets). The explosive-assembly makeup room is environmentally controlled. Three firing stations are maintained in an operational mode.

The costs assumed for the facility are listed below:

- Building (15,000 ft)^2 = $600,000
- Equipment and furnishings = $750,000
- Land (and site preparation) = $200,000.

5.   Facility Staffing and Salary Bases for Cost Estimate

The facility staffing includes production workers, engineering and quality assurance personnel, supervisors, administrative personnel, and other support personnel. Additional services are supplied by the parent company at an annual charge (which is based on level of business) or are included in the division overhead.

The staff members are identified in Table 19. The production workers listed in the table are matched to specific process operations listed in Figure 21.

TABLE 19
FACILITY STAFFING ASSUMPTIONS

1.   Production Workers
   a.   Firing Assembly Crews [Operations 8 through 11 in Fig. 21]
      ● Super Technician (2)
      ● Technician (2)
   b.   Compaction Assembly and Support Crews [Operations 1 through 7 and 12 in Fig. 21]
      ● Senior Technician (2)
      ● Technician (4)
      ● Helper (2)

2.   Quality Assurance Staff [Incoming materials qualification and inspection, process surveillance, product qualification and testing]
      ● Engineer (1)
      ● Senior Inspector (1)

3.   Engineering Staff [Process engineering and maintenance]
      ● Senior Engineer (1)
      ● Senior Technician (1)

4.   Administrative Support Staff
      ● Secretary-Administrative Assistant (1)
      ● Junior Engineer (1) [Production scheduling and control]

5.   Management Staff
      ● Plant Manager (1)
      ● Shift Supervisor (2)

Annual salaries are listed in Table 20 for each employee grade listed in Table 19. The total annual salary is $631,000, of which $282,000 (45 percent) is attributable to production workers. The division of salaries between shifts is 71 percent for the day shift and 29 percent for the night shift.

TABLE 20

ANNUAL SALARIES ASSUMED FOR STAFF MEMBERS

	Staff Members	Salary	Subtotal
1.	Production Workers		
	• Super technician (2)	$ 35,000	$ 70,000
	• Senior technician (2)	32,000	64,000
	• Technician (6)	20,000	120,000
	• Helper (2)	14,000	28,000
		Subtotal	$282,000
2.	Quality Assurance Staff		
	• Engineer (1)	$ 40,000	$ 40,000
	• Senior inspector (1)	32,000	32,000
		Subtotal	$ 72,000
3.	Engineering Staff		
	• Senior engineer (1)	$ 45,000	$ 45,000
	• Senior technician (1)	32,000	32,000
		Subtotal	$ 77,000
4.	Administrative Support Staff		
	• Secretary-admin. assistant (1)	$ 30,000	$ 30,000
	• Junior engineer (1)	30,000	30,000
		Subtotal	$ 60,000
5.	Management Staff		
	• Plant manager (1)	$ 60,000	$ 60,000
	• Shift supervisor (2)	40,000	80,000
		Subtotal	$140,000
		TOTAL	$631,000

6.   Estimate of Total Annual Cost

As shown in Table 21, the total annual cost for operation and support of the dynamic compaction facility is about $10 million. This total includes the direct labor costs from Table 20 and the direct material costs from Table 18. Allowances are made in Table 21 for corporate support (Item B) and for division overhead (Item C), which are derived from direct costs (Item A). The total cost also includes capital depreciation allowances for building and equipment (Item D). The G&A allowance (Item E) is derived from the direct costs (Item A) and overhead (Item C). Allowances are also made for the cost of money (Item F), including payment for land.

The total annual cost is heavily dependent on the cost of process materials (74 percent). Within the materials costs, TiB_2 powder is the most dominant factor (about 47 percent of total annual costs).

The total annual cost equates to $34.70 per pound of TiB_2 that is contained in the shipped product. Thus, the value-added to the original ceramic powder during dynamic processing to final product form is 1.3 times ([$34.70 ÷ 15] - 1) in respect to cost.

7.   Product Pricing Estimate

The pricing estimate for the product is based on an assumption that the parent corporation has no large tax writeoffs. The total annual revenue is obtained by providing for a 17.5 percent profit margin over costs and a tax allowance of 48 percent over cost and profit.

The total annual revenue value is $23,396,300 ([$10,037,000 cost]) ÷ [1 - 0.175] ÷ [1 - 0.48]). This total revenue equates to a unit value of approximately $81 per pound of TiB_2 contained in the shipped product.

TABLE 21

ESTIMATION OF TOTAL ANNUAL COST FOR
OPERATION OF THE DYNAMIC COMPACTION FACILITY

A.	Direct Costs	$ 8,009,400	(80%)
	1. Labor (Table 20)	$ 631,000	(3%)
	2. Process materials: $392.20 per firing (Table 18) for 18,813 firings	$ 7,378,400	(74%)
B.	Corporate Support	$ 204,400	(2%)
	1. Research and development: 2% of direct costs	160,200	
	2. Administrative (including legal and procurement: 7 percent of direct labor costs)	44,200	
C.	Division Overhead (including fringes, utilities, plant and office supplies, janitorial and security services: 50 percent of direct labor costs)	$ 315,500	(3%)
D.	Capital Depreciation	$ 180,000	(2%)
	1. On building (15,000 sq ft; cost = $600,000): 5% per year	30,000	
	2. On equipment (including magazines and firing closets): cost = $750,000: 20% per year	150,000	
E.	General and Administrative: 7% of direct costs and division overhead	$ 582,700	(6%)
F.	Interest	$ 745,000	(7%)
	1. On Items A, B, C, and E, above: 8%	729,000	
	2. On land (200 acres; cost = $200,000): 8%	16,000	
G.	Total Annual Cost	$10,037,000	(100%)

C.   **BASE COST AND PRICE ESTIMATES FOR SHS AND DENSIFICATION OF CERAMIC TILES**

1.   **Product Description**

The product shape--a flat tile, shown in Figure 23--was selected as a simple geometric configuration that could be produced by SHS technology and could be used in ceramic/metal structural assemblies. Although the tile is amenable to fabrication by SHS technology, its dimensions would pose considerable difficulty for fabrication by standard hot pressing on a hydraulic or mechanical press. Assuming a density of 98 percent in TiB_2, the annual production requirement relates to approximately 50,000 tiles.

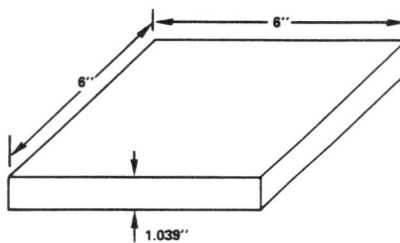

MATERIAL: TiB_2
WEIGHT: 2.69 kg (5.94 lb), AT 98% DENSITY

NOTE: NOT-TO-SCALE

FIGURE 23

CERAMIC TILE FORMED BY SHS TECHNOLOGY

Product density of more than 98 percent should be obtainable by the SHS process that is described in Section 3. In addition, machining of the reacted tile is assumed to be unnecessary. Since the reacting material is at least partially liquid during SHS processing, control of dimensions should be comparable to that achieved by precision casting.

2.   SHS Process Basis for Cost Estimate

The basic assumption is that 21 tiles are pressed from a process blend of titanium and boron powders and simultaneously reacted and densified in SHS reaction chambers. The operational sequence for a powder blend is shown in the flow diagram in Figure 24 and summarized below:

- Preparation of the powder blend (Steps 1-5)
- Pressing and preparation of the green TiB_2 tiles for SHS reaction (Steps 6-9)
- SHS reaction, densification, and cooling (Steps 10-13).

The process operations and parameters are based on various descriptions of SHS technology in Soviet literature, but the greatest reliance has been placed on a Soviet patent [Ref. 38]. That patent describes a process for manufacturing ceramic (titanium carbide) tool bits. The pressing setup has been modified slightly for producing tiles.

The Soviet tool bits are prepressed as green powder compacts in a die and subsequently reacted and densified between molybdenum electrodes in a reaction chamber in which an inert atmosphere is pressurized between 1/2 and 5 atmospheres. For this estimate, a highly automated line is assumed--with operator access to the work through gloveports in work station enclosures. Small reaction chambers, which hold individual tiles, are envisioned inside the glovebox enclosures.

Although a large press (750 tons) is required to reach the pressure levels reported in Soviet literature for green pressing, much larger units have been made for the domestic powder metallurgy industry. One alternative would be to use a smaller press to produce smaller tiles. For example, 4-inch square tiles would require approximately one-half of the press capacity needed for 6-inch square tiles. Another alternative would be to determine powder characteristics that would provide the same green density at a lower pressure.

OPERATIONS[a]	WORK UNITS		
	POWDER PREPARATION	PRE-REACTION TILES	REACTED TILES
1. RECEIVING STORAGE	TITANIUM AND BORON POWDERS		
2. V-BLEND POWDERS (1 hr)	POWDER BLEND (57 kg)		
3. WET BALL MILL BLEND (6 hr)	MILLED BLEND		
4. DRY POWDER BLEND (2 hr)	DRIED BLEND		
5. STORE POWDER BLEND	DRIED BLEND		
6. PRESS GREEN TILES (3,000 kg/cm^2)		GREEN TILES (60% DENSE)	
7. ADD THIN LAYER OF DRIED POWDER BLEND ON IGNITION SURFACE OF TILES		COATED GREEN TILES	
8. PRESS COATED TILE (500 kg/cm^2)		PRE-REACTION TILES	
		PRE-REACTION TILES (21 pcs)	
9. STORE UNDER VACUUM (10^{-2}mm Hg)			REACTION LOADING[b]
10. LOAD TILE BETWEEN REACTION ELECTRODES			TILE
11. PREHEAT TILE AND IGNITE[c] (WITH ELECTRICAL CURRENT PASSING BETWEEN ELECTRODES)			
12. HOLD THROUGHOUT SHS REACTION[c]			
13. COOL AND UNLOAD TILE			REACTED DENSE TILES (21)

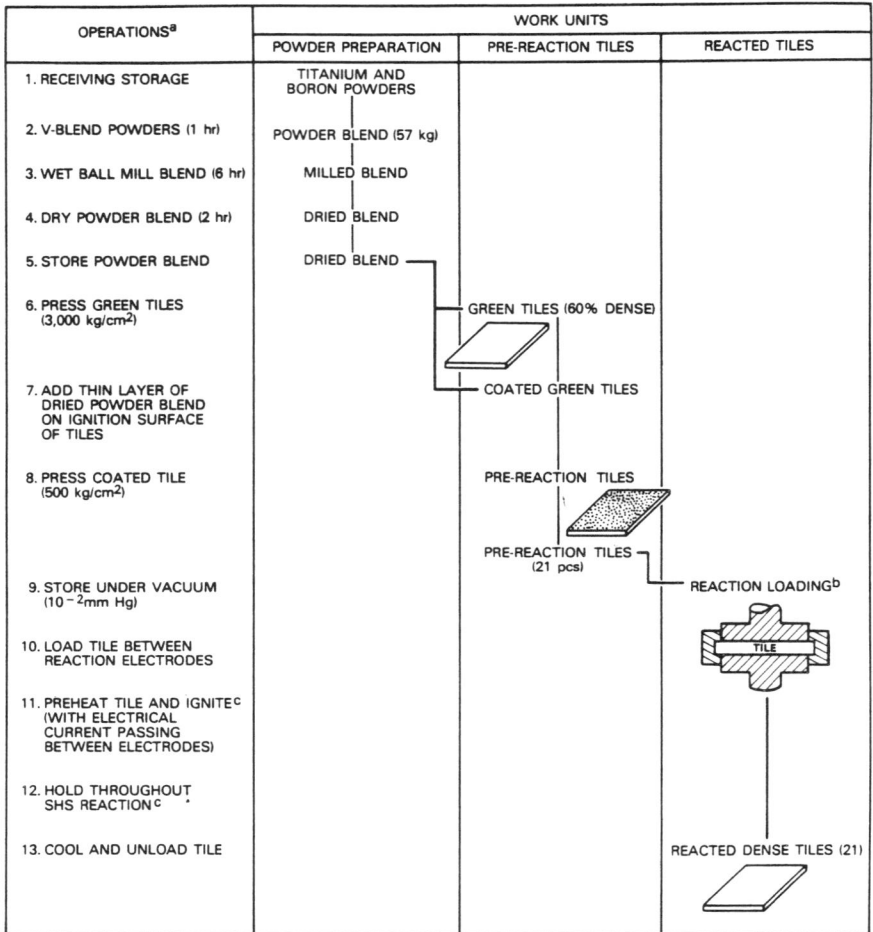

[a] UNDER INERT ATMOSPHERE (OR VACUUM, WHEN SPECIFIED).
[b] SEE FIGURE 15 FOR MORE DETAILS.
[c] PRESSURE OF INERT ATMOSPHERE INCREASED TO 5 ATM.

FIGURE 24

PROCESSING OPERATIONS FOR SHS REACTION
AND DENSIFICATION OF TiB$_2$ TILES

The SHS reaction is ignited by an electrical current that is passed between the two electrodes. Ignition is aided by prepressing some loose powder into the ignition surface at a low pressure level.

The TiB_2 tile should be partially or fully molten during the SHS reaction. Precision dimensions are maintained by squeezing the reaction mass between the electrodes within a refractory retainer as shown in Figure 15. This squeeze could be provided by a small hydraulic or mechanical press, by a spring, or by a dead weight mechanism.

3. Process Materials Basis for Cost Estimate

The raw materials are assumed to be titanium sponge and amorphous boron powder. Both materials have been used in the U.S.S.R. to produce titanium diboride by SHS. The unit costs of those powders, shown in Table 22, are consistent with the value of $15.00 per pound for titanium diboride powder that was used in the previous cost analysis for dynamic compaction of TiB_2 disks. These values are between 30 and 45 percent of current prices for the same powders when purchased in small quantities (i.e, 100 pounds or less).

TABLE 22
MATERIAL COSTS PER 57-KG POWDER BLEND

1. Titanium (sponge; approximately 60-mesh particle size; 0.3 percent oxygen, maximum)

   $4.25/lb x 57 kg x 2.2 lb/kg x 0.689 (mass percent) = $367

2. Amorphous Boron (2- to 3-micron particle size; 92 to 95 percent purity)

   $27.50/lb x 57 kg x 2.2 lb/kg x 0.311 (mass percent) = $1,073

3. Total Cost = $1,440/blend

As shown in the table, the cost of powder per processing blend is $1,440. Approximately 75 percent of this value is attributable to the cost of boron. and the remainder represents the cost of titanium.

4. Facililty Basis for Cost Estimate

The facility is assumed to be free standing and of "Butler-type" construction. The building contains standard powder processing and inspection equipment. The production lines are mostly housed in gloveboxes that are filled with inert gas.

Five powder blends, approximately 105 tiles, are processed per shift. At an assumed worker productivity level of 70 percent, 7 operator stations and a total of 21 SHS reactors are estimated to be required.

The costs assumed for the facility are listed below:

- Building (12,600 ft^2) = $ 504,000
- Equipment and furnishings = $1,120,000
- Land (10 acres) and site preparation = $30,000.

5. Facility Staffing and Salary Bases for Cost Estimate

The facility staffing includes production workers, engineering and quality assurance personnel, supervisors, administrative personnel, and other support. Additional services are supplied by the parent company at an annual charge (which is based on level of business) or are included in the division overhead.

The staff members are identified in Table 23. The production workers listed in the table are matched to specific process operations listed in Figure 24.

TABLE 23

FACILITY STAFFING ASSUMPTIONS

1. Production Workers

   a. Blend preparation crews [Operations 1-5]
      - Senior technician (2)
      - Technician (2)

   b. Green tile crews [Operations 6-9]
      - Senior technician (2)

   c. SHS reaction and densification [Operations 10-13]
      - Super technician (2)
      - Senior technician (4)
      - Technician (6)
      - Junior technician (2)

2. Quality Assurance Staff [Incoming materials qualification and inspection, process surveillance, product qualification and testing]
      - Engineer (1)
      - Senior inspector (1)
      - Junior inspector (1)

3. Engineering Staff [Process engineering and maintenance]
      - Senior engineer (1)
      - Senior technician (1)
      - Junior technician (1)

4. Administrative Support Staff
      - Secretary-administrative assistant (1)
      - Junior engineer (1) [Production scheduling and control]

5. Management Staff
      - Plant manager (1)
      - Production superintendent (1)
      - Night supervisor (1)
      - Production foreman (2)

Annual salaries are listed in Table 24 for each employee grade listed in Table 23. The total annual salary is $977,000, of which $514,000 (53 percent) is attributable to production workers. The division of salaries between shifts is 62 percent for the day shift and 38 percent for the night shift.

TABLE 24

ANNUAL SALARIES ASSUMED FOR STAFF MEMBERS

Staff Members	Salary	Subtotal
1. Production Workers		
• Super technician (2)	$ 35,000	$ 70,000
• Senior technician (8)	32,000	256,000
• Technician (8)	20,000	160,000
• Junior technician (2)	14,000	28,000
	Subtotal	$514,000
2. Quality Assurance Staff		
• Engineer (1)	$ 40,000	$ 40,000
• Senior inspector (1)	32,000	32,000
• Junior inspector (1)	17,000	17,000
	Subtotal	$ 89,000
3. Engineering Staff		
• Senior engineer (1)	$ 45,000	$ 45,000
• Senior technician (1)	32,000	32,000
• Junior technician (1)	17,000	17,000
	Subtotal	$ 94,000
4. Administrative Support Staff		
• Secretary-admin. assistant (1)	$ 30,000	$ 30,000
• Junior engineer (1)	30,000	30,000
	Subtotal	$ 60,000
5. Management Staff		
• Plant manager (1)	$ 60,000	$ 60,000
• Production superintendent (1)	45,000	45,000
• Night supervisor (1)	40,000	40,000
• Production foreman (2)	37,500	75,000
	Subtotal	$220,000
	TOTAL	$977,000

6.   Estimate of Total Annual Cost

As shown in Table 25, the total annual cost for operation and support of the SHS facility is between $6 and 7 million. This includes the direct labor costs from Table 24 and the direct material costs from Table 22. The allowances for corporate support (Item B), division overhead (Item C), capital depreciation (Item D), G&A (Item E), and interest (Item F) were calculated as described previously for values listed in Table 21 in regard to dynamic compaction of TiB_2 disks.

TABLE 25

ESTIMATION OF TOTAL ANNUAL COST FOR
OPERATION OF THE DYNAMIC COMPACTION FACILITY

A.	Direct Costs	$ 4,577,000	(73%)
	1. Labor (Table 24)	$ 977,000	(16%)
	2. Process materials: $1,440 per powder blend (Table 22) for 2,500 blends	$ 3,600,000	(57%)
B.	Corporate Support	$ 159,000	(2%)
	1. Research and development: 2% of direct costs	91,500	
	2. Administrative (including legal and procurement: 7 percent of direct labor costs)	68,400	
C.	Division Overhead (including fringes, utilities, plant and office supplies, janitorial and security services: 50 percent of direct labor costs)	$ 488,500	(8%)
D.	Capital Depreciation	$ 249,200	(4%)
	1. On building (12,600 sq ft; cost = $504,000): 5% per year	25,200	
	2. On equipment (including press, etc.): cost = $1,120,000: 20% per year	224,000	
E.	General and Administrative: 7% of direct costs and division overhead	$ 354;600	(6%)
F.	Interest	$ 448,800	(7%)
	1. On Items A, B, C, and E, above: 8%	446,400	
	2. On land (10 acres; cost = $30,000): 8%	2,400	
G.	Total Annual Cost	$ 6,278,000	(100%)

The total annual cost is heavily dependent on the cost of process materials (57 percent). Within the materials costs, boron powder is the most dominant factor (almost 45 percent of total annual costs).

The total annual cost equates to $22 per pound of TiB_2 that is contained in the shipped product. Thus, the value-added to the original ceramic powder during SHS processing to final product form is 0.9 times ([$22 ÷ 11.56] - 1) in respect to costs.

7.   Product Pricing Estimate

The pricing estimate for the product is based on an assumption that the parent corporation has no large tax writeoffs. The total annual revenue is obtained by providing for a 17.5 percent profit margin over costs and tax allowance of 48 percent over cost and profit.

The total annual revenue value is $14,634,000 ([$6,278,000 cost]) ÷ [1 - 0.175] ÷ [1 - 0.48]). This total revenue equates to a unit value of approximately $51 per pound of TiB_2 contained in the shipped product.

D.   PRICING COMPARISONS FOR TITANIUM DIBORIDE SHAPES MANUFACTURED BY DCT AND SHS TECHNOLOGY

1.   Comparison With Existing Production Technology

Hot-pressed flat tiles of TiB_2 have sold in small sizes and in small quantities over the past few years for per-pound values in the range of $100 to $400. Recently, a value was reported, but not confirmed, at a level of $75 to $80 per pound. Thus, based on these approximations of $81 and $51 per pound, the new technologies should be cost effective in comparison with existing production technology.

2.   Comparison of DCT and SHS Technology

The major factor contributing to the fairly large percentage difference [59 percent = ($81 - 51) ÷ $51] between the base prices for processing TiB_2 metal-jacketed disks and bare tiles is the difference in cost of process materials [$7,378,400 (Table 21) vs $3,600,000 (Table 25)]. Whereas

TiB_2 powder and other process materials were purchased for DCT processing, only elemental titanium and boron powders were required for SHS processing. The SHS combination of synthesis and densification operations provides a considerable financial advantage over DCT and other two-step processes. This cost advantage for SHS processing overrides the considerably higher estimates in Tables 25 and 21 for labor ($977,000 vs $631,000) and for capital expenditure for equipment ($1,120,000 vs $750,000) in comparison to the DCT process.

E.   USE OF EXPERIENCE CURVES IN PREDICTING PRICE BEHAVIOR

1.   Methodology

The forecast of pricing behavior beyond the time period of the base cost was developed by use of the experience (or learning) curve technique. Since manufacturers often must make such projections without much factual information, reliance is placed on similar experiences. These experiences have been recorded in the form of curves that relate either cost or price behavior to cumulative production quantity.

Although the economic theory behind experience curves is rather complex [Ref. 73], the principle can be stated simply: *price-per-unit declines by some constant percentage as the quantity of units produced doubles*. When this relationship is illustrated on a log-log plot, it is linear, as shown in Figure 25 for three possible levels of learning experiences. To illustrate for the 90 percent curve in the figure, the initial price of each unit has decreased by a factor of 0.9 times the prior value (100 percent) when the cumulative production quantity has increased from one to two units. In this case, the initial unit price will have decreased more than one-half at a production level considerably less than 1,000 units.

The experience curve concept has been used in this study to demonstrate how product prices would be expected to decline when advanced processing technologies are used in a production environment over a period of years. The base (or starting points) are the estimated prices for ceramic shapes that were developed in Sections B and C.

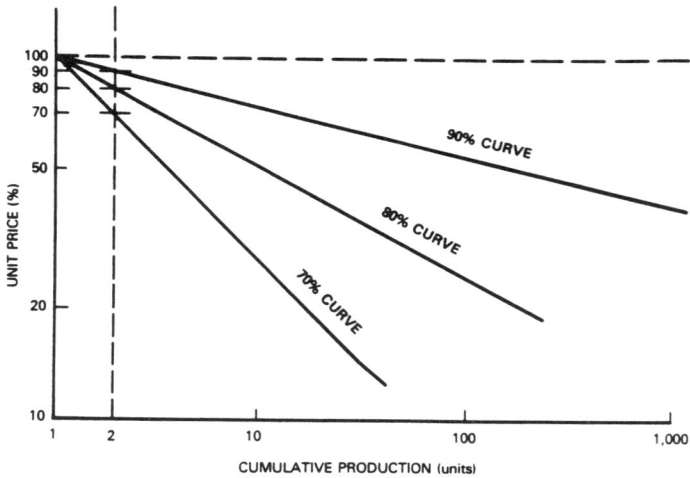

FIGURE 25

ILLUSTRATION OF 70, 80, AND 90 PERCENT EXPERIENCE CURVES

2.   Background

The experience curve concept is applicable to any product to which a production process contributes added value. Figures 26 and 27 illustrate actual market behavior for two entirely different types of product--an electronic component and a plastic raw material. In Figure 26, the relatively steep 70 percent experience curve and the relatively rapid doublings of production (between five and six times within a span of 3 years) reflect the strong demand and intense production competition for integrated circuits during the mid-1960s. Both annual (dots) and monthly (diamonds) unit prices conform rather closely to the curve that has a 70 percent slope. In the case of low-density polyethylene in Figure 26, only four market doublings occurred within a span of 7 years (1952-59), and unit prices followed a 90 percent experience curve during that period. However, market factors caused a drastic change in the experience curve for this product during the 1960s: the slope changed abruptly, from 90 percent to approximately 65 percent. This abrupt change in slope occurred without a drastic change in demand, since it was accompanied by only two to three doublings over this period of 9 years. Competitive influences in the market and major improvements in processing technologies are likely causes for the change in slope.

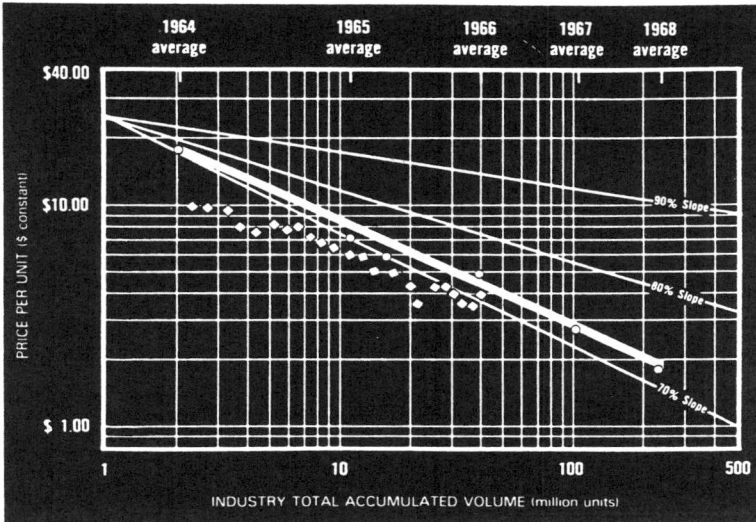

O Average annual prices in constant dollar value.
◇ Average monthly prices in constant dollar value.

SOURCE: Reference 73.

FIGURE 26

EXPERIENCE CURVES PLOTTED FOR PRICE
BEHAVIOR OF INTEGRATED CIRCUITS DURING MID-1960s

Although raw data are not available for plotting an experience curve for prices for zirconium products, a situation similar to the ones illustrated in Figures 26 and 27 regarding price behavior of this metal was noted during the 1950s and 1960s in the naval and commercial nuclear power industries. Probably only a few pounds of refined zirconium metal were

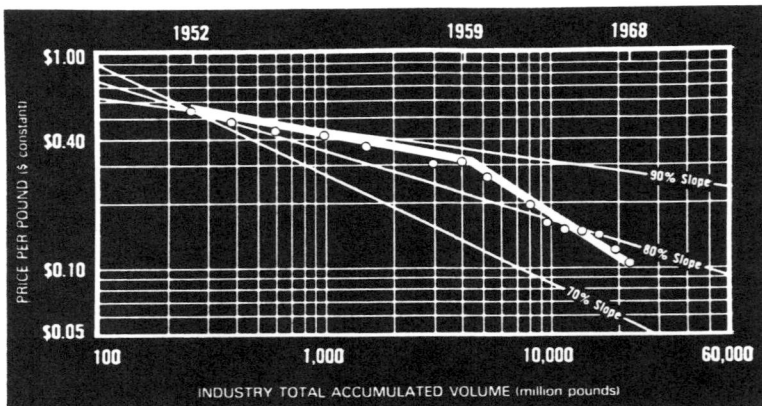

O Average annual prices in constant dollar value.

SOURCE: Reference 73.

FIGURE 27

**EXPERIENCE CURVES PLOTTED FOR PRICE
BEHAVIOR OF LOW-DENSITY POLYETHYLENE DURING 1950s AND 1960s**

available in the United States in the late 1940s. For technical reasons, the metal was selected as the fuel cladding and fuel assembly structural material in early nuclear power reactors. Whereas zirconium components initially had cost thousands of dollars on a per-pound basis of contained zirconium, within a period of 10 to 15 years this unit price decreased to tens of dollars or less for essentially all standardized reactor components. During this period, two naval and two commercial reactor prototypes were constructed, and a number of naval production reactors were completed; the cumulative production of zirconium products increased from a few pounds to hundreds of tons.

The actual price behavior of titanium sponge probably paralleled the situation described for zirconium reactor components except that the value-added during production of sponge is considerably less than for the finished zirconium products. The pricing data shown in Figure 28 followed an experience curve in the range of 95 to 100 percent during the early 1950s as cumulative production of titanium sponge increased from 55 to 11,000

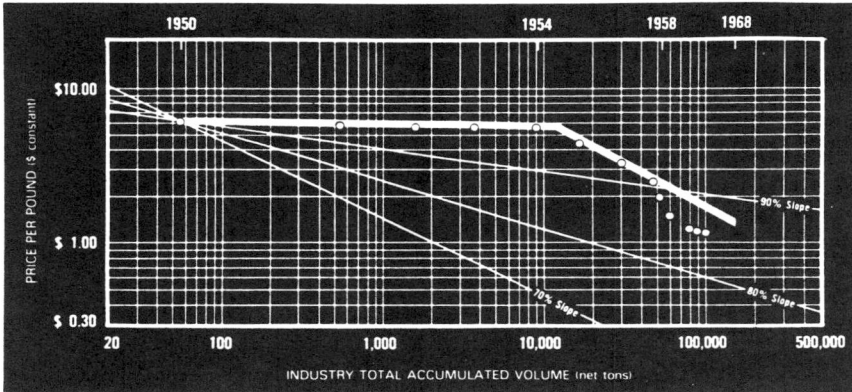

○ Average annual prices in constant dollar value.

SOURCE: Reference 73.

FIGURE 28

EXPERIENCE CURVES PLOTTED FOR PRICE
BEHAVIOR OF TITANIUM SPONGE DURING 1950s AND 1960s

tons by 1954, when annual production was approximately 7,000 tons. During the late 1950s and 1960s, as annual production increased to between 15,000 and 20,000 tons, the pricing behavior followed an experience curve with a slope that was somewhat steeper than 70 percent.

The pricing behavior described for zirconium and titanium, especially in the case of finished zirconium products, is not unlike that which would be expected for the ceramic shapes described in Sections B and C. This demonstration reveals clearly what would be expected to happen to the current relatively high prices of ceramic tiles after introduction into a major production program. The selection of specific experience curves used in the demonstration is broad enough to suggest that the extreme values shown probably are representative of maximum and minimum prices that could result; also, a most likely behavior has been deduced within the broad range.

F.   EXPERIENCE CURVES APPLIED TO PRODUCTION OF CERAMIC SHAPES BY DCT AND SHS TECHNOLOGY

1.   Calculational Basis

Potential structural applications of ceramic/metal assemblies have been assumed for both defense and industrial markets. In addition, a considerable industrial follow-on market for similar dense TiB_2 products is anticipated. For example, an application in aluminum smelting electrodes has been assumed in the time interval between the structural applications. This market might reach millions of pounds per year in the United States alone. Overall, the assumptions are quite conservative in regard to potential demand for the assumed products. Many potential uses have been ignored, and the production rates for the assumed uses are quite modest.

Table 26 presents the annual mass quantities of TiB_2 for these three applications and the annual totals for a 15-year period, which has been divided into an industry startup phase (years -4 through -1) and a production phase (years 1 through 11). As indicated, the annual production level of 1 ton in the first year is increased slowly to values over 1,000 tons. The cumulative quantity builds over this period from a starting level of a few tons to almost 15,000 tons for these applications. As shown at the bottom of the table, six doublings occur during the 11-year production phase.

The other need in assessing price behavior is a starting value for the unit price of TiB_2. The values of $81 and $51 per pound that were developed previously (Sections B.7 and C.7) were used for this purpose in the first production year, when the production rate is assumed to be 385 structural assemblies.

2.   Construction of Experience Curves for DCT

The production quantities in Table 26 and the base pricing value for dynamic compaction of TiB_2 disks were used to construct the four experience curves in Figure 29. Three curves are plotted for individual learning expectations of 90, 80, and 70 percent; also, a duplex curve (marked by the

TABLE 26

TABULATION OF TiB_2 PRODUCTION TONNAGE

Applications[a]	Startup Phase Years				End-of-Year Tons of TiB_2 (1,000 kg each) — Production Phase Years										
	-4	-3	-2	-1	1	2	3	4	5	6	7	8	9	10	11
First Assembly Type															
• Development	1	1.5	2.00	1											
• Prototypes			1.25	4											
• Production					131	380	380	380	380	380	380	380	380		
Other Industrial use															
• Prototypes				1	3	1									
• Production						5	20	30	40	50	60	72	84	96	115
Second Assembly Type															
• Development & test					5	2	3	10							
• Prototypes						3									
• Production									500	1,000	1,800	1,800	1,800	1,800	1,800
Total TiB_2															
• Annual	1	1.5	3.25	6	139	390	403	420	920	1,430	2,240	2,252	2,264	1,896	1,915
• Cumulative	6[a]	7.5	11.00	17	156	546	949	1,369	2,289	3,719	5,959	8,211	10,475	12,371	14,286
Cumulative Production Doublings					*	* (1)	* (2)	* (3)		* (4)	* (5)		* (6)		

[a] Includes an estimate of prior cumulative TiB_2 production of 5 tons.

heavy line) is plotted for an initial learning expectation of 85 percent, followed by a steeper slope that represents a learning expectation of 75 percent. The 90 and 70 percent curves are considered to be limiting values for this situation, and the duplex curve represents the preference for the most likely price behavior.

The curves have been constructed backward (dotted lines) and forward (solid lines) from the initial production pricing estimate of $81 per pound of TiB_2 at 156 tons of cumulative production. When extrapolated backward to year -4, the price range varies between the extremes of $125 and $345. This range represents prices for TiB_2 development disks that are similar to ceramic tiles sold in small quantities over the past few years (refer to Section D.1).

The initial slope in the duplex curve (85 percent) was changed (to 75 percent) when the DCT operations were moved from development facilities to a production plant. The change in slope follows four fairly rapid doublings of cumulative production that occurred during years -1 and -2; in addition, the jump in annual production from 6 to 139 tons of TiB_2 necessitates the change to new facilities.

The initial slope is a compromise between the slowest dropping curve and middle curve (90 and 80 percent, respectively) among the other three; the slope in later years is a compromise between the steepest dropping curve and middle curve (70 and 80 percent, respectively). Selection of this duplex curve as the most probable expectation for behavior of prices has been influenced to a great degree by the actual experience recorded in Figures 27 and 28.

Pricing data from all four experience curves in Figure 29 are listed in Table 27, in constant 1985 dollar values, for the 6th and 11th years into the production phase. As expected, the full pricing range is quite broad; the unit values in the 11th year range from $8 to $39 per pound of contained TiB_2. Thus, the lowest value is approximately one-tenth of the starting production value ($81 per pound of TiB_2), and the highest value is approximately one-half of the initial value. The preferred pricing value, in the 11th production year, is $12 per pound of contained TiB_2.

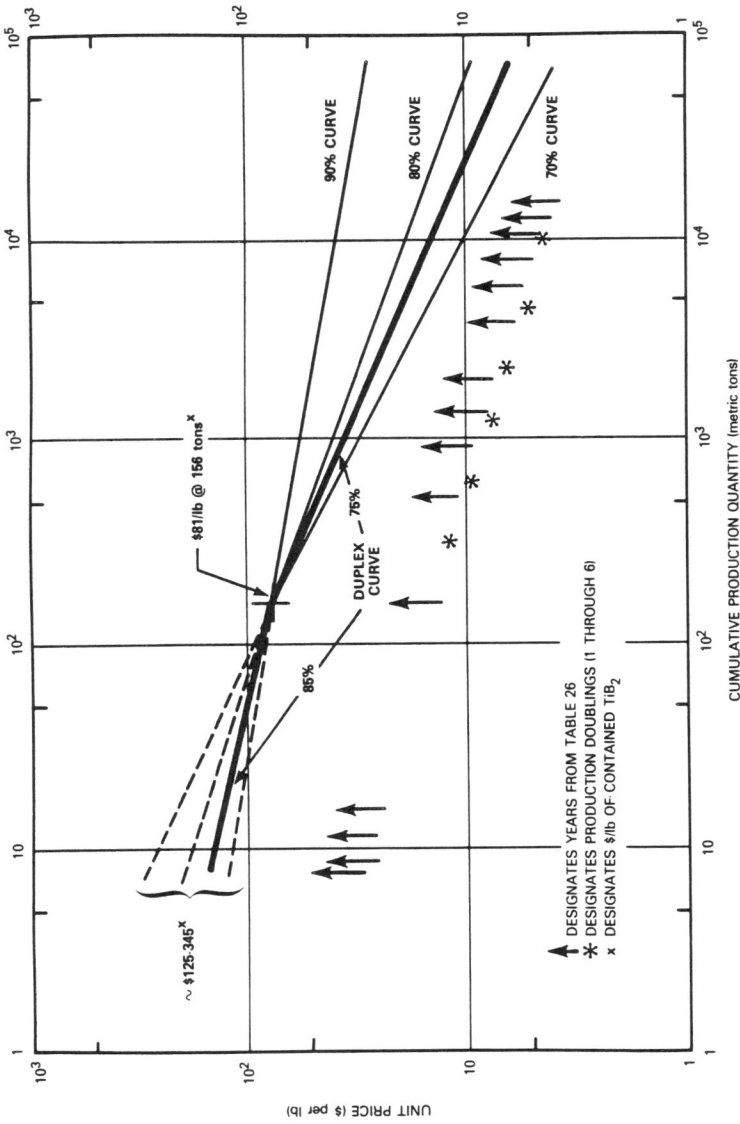

FIGURE 29

A GROUP OF EXPERIENCE CURVES USED TO DEMONSTRATE PRICE BEHAVIOR
OF TiB_2 ANTICIPATED OVER A 15-YEAR PERIOD

TABLE 27

CHANGES ANTICIPATED IN TiB$_2$ PRICES AFTER 11 YEARS OF PRODUCTION

| Years of Production | TiB$_2$ Price (\$ per lb)[a] for Various Experience Curves | | | |
	70%	80%	90%	Duplex (85/75%)
6th year	15	29	50	21
11th year	8	19	39	12

[a]Constant dollar values.

3. Extended Pricing Values for SHS Technology

When the same forecasting technique and production assumptions are used to extend the base price of TiB$_2$ tiles produced by SHS technology (\$51 per pound), a similar projection is obtained relatively easily. In this instance, the most likely projection obtained per pound of contained TiB$_2$ in the tiles is \$8 after the 11th production year. Similarly, the high and low projections for that same year are \$26 and \$6, respectively.

References

1. Outlook for Science and Technology: The Next Five Years, National Research Council, W. H. Freeman and Co., 1982.

2. Trends and Opportunities in Materials Research, Materials Research Advisory Committee, National Science Foundation, 1984.

3. Materials and Structures Technology Conference, June 14-16, 1983, The National Materials Advisory Board of the National Research Council, June 1983.

4. ASM Conference on Materials for Future Energy Systems, 1-3 May 1984, American Society for Metals Technical Divisions Board, May 1984.

5. Dr. Dale E. Niesz et al., "Materials," Research and Development, June 1984, pp. 266-72.

6. Harry E. Chandler, "Emerging Trends in Aerospace Materials and Processes," Metal Progress, April 1984, pp. 21-29.

7. Dr. Raymond L. Smith, "Atlas Never Shrugged," ASM News, February 1984, pp. 3 and 6.

8. High-Temperature Metal and Ceramic Matrix Composites for Oxidizing Atmosphere Applications, NMAB-376, National Materials Advisory Board, Commission on Sociotechnical Systems, 1981.

9. "Technology Forecast, '83," Metal Progress, January 1983, pp. 14-80.

10. Proceedings of the Ninth DARPA Strategic Space Symposium, Vols. I and II, 4-7 October 1983, DARPA-TIO-84-1, Defense Advanced Research Projects Agency, January 1984 (classified Secret/Formerly Restricted Data).

11. First Symposium on Space Nuclear Power Systems, 11-13 January 1984, at The University of New Mexico, Orbit Book Company, Inc., 1985.

12. Refractory Alloy Technology for Space Nuclear Power Applications, CONF-8308130, TIC, U.S. Department of Energy, January 1984.

13.   V. L. Ginzburg, "What Special Problems of Physics and Astrophysics Are of Special Importance and Interest at Present? (Ten years later)," Soviet Physics. Uspekhi, Vol. 24, No. 7, July 1981, pp. 585-613.

14.   W. L. Frankhouser et al., Gasless Combustion Synthesis of Refractory Compounds, Noyes Publications, 1985.

15.   "Powers That 'Explode' Into Materials," Advanced Materials & Processes, Vol. 2, Issue 2, February 1986, pp. 25-32.

16.   A Competitive Assessment of the U.S. Advanced Ceramics Industry, PB84-162288, U.S. Department of Commerce, March 1984.

17.   1985 U.S. Industrial Outlook, Prospects for Over 850 Industries, U.S. Department of Commerce, January 1985.

18.   "Advanced Ceramics, The U.S. Edge Is in Danger," Chemical Week, 6 February 1985, pp. 38-42.

19.   High Technology Ceramics in Japan, NMAB-418, National Academy Press, 1984.

20.   Garret/Ford AGT 101 Advanced Gas Turbine Program Summary, Garret Turbine Engine Company, June 1984.

21.   "Overview of the German Ceramic Gas Turbine Program," Ceramics for High-Performance Applications, III, Reliability, Vol. 6, Plenum Press, 1979, pp. 29-50.

22.   "Ceramic Engine Research and Development in Sweden," Ceramics for High-Performance Applications, III, Reliability, Vol. 6, Plenum Press, 1979, pp. 51-79.

23.   Ceramic Roller Bearing Development Program Phase II, Phase III Report, NAPC-PE-95C, Teledyne CAE, June 1984.

24.   Peter A. Jung, Composite Periscope Barrel, Technical Report 939, Naval Ocean System Center, January 1984.

25.   "Advanced Ceramic Combustor/Heat Exchanger Engine System," Proceedings: Workshop on Ceramics for Advanced Heat Engines, CONF-770110, Energy Research and Development Administration, 1977, pp. 89-113.

26.   E. S. Atroshenko et al., "The Effect of Explosive Compacting on the Properties of 60% $LaCrO_3$ 40% Cr Cermet," High-Temperatures, High-Pressures, Vol. 8, 1976, pp. 21-26.

27.   A. A. Deribas and A. M. Staver, "Shock Compression of TiO_2 + $BaCo_3$ Powder Mixture," Fizika Goreniya i Vzrvya, Vol. 6, No. 1, 1979, pp. 122-23.

28.   Fine Ceramics, Kyoto Ceramic Co., Ltd., Advertising Literature, circa 1980.

29.   A. N. Kiselev, "Explosion Forming of $SmCo_5$," Fizika Goreniya i Vzryva, Vol. 13, No. 1, January-February 1977, pp. 118-20.

30.   Technological and Economic Assessment of Advanced Ceramic Materials, Volume 6, "A Case Study of Ceramic Cutting Tools," PB85-113132, National Bureau of Standards, August 1984, p. 66.

31.   James Dickson, Rapid Solidification Technology Program Review, SPC 857, System Planning Corporation, October 1982.

32.   V. P. Khlyntsev and L. P. Potapov, "Study of the Structure and Crystallization of Metallic Glass $Ni_{70}Ta_{30}$," Metallofizika, No. 1, 1980.

33.   "Study of the Crystallization of Metallic Glass $Ni_{60}Nb_{40}$," Soviet Physics of Metals and Metallography, Vol. 51, No. 6, 1981, pp. 85-89.

34.   O. V. Roman et al., "Structure and Properties of an Amorphous Powder Material After Explosive Loading," Metal Science and Heat Treatment, No. 10, 1984, pp. 781-84.

35.   Y. Y. Kalinin, "Magnetoelastic Decay and the WE-Effect in an Amorphous Fe-Ni Alloy," Physics of Metals and Metallography, Vol. 55, No. 2, May 1984, pp. 29-33.

36.   C. C. Kock et al., "High-Temperature Ordered Intermetallic Alloys," Materials Research Society Proceedings, Materials Research Society, Vol. 39, 1985.

37.   A. G. Merzhanov et al., "Method for Production of Two-Layer Pipe Casting," U.S. Patent No. 4,217,948, 18 August 1980.

38.   A. G. Merzhanov et al., "Method for Obtaining Refractory Compounds," U.S.S.R. Patent No. 584,052, 15 December 1977.

39.   V. I. Itin et al., "Structure and Properties of Titanium Nickelide Materials Made by Self-Propagating High-Temperature Synthesis," Izvestiya Vysshikh Uchebnykh Zavedenii Fizika, No. 12, 1977, pp. 117-20.

40.   A. D. Bratchikov, "Self-Propagating High-Temperature Synthesis of Titanium Nickelide," Poroshkovaya Metallurgiya, No. 1, pp. 7-11, 1980.

41.   V. I. Itin et al., "Production of Titanium Nickelide by Self-Propagating High-Temperature Synthesis," Poroshkovaya Metallurgiya, No. 3, pp. 4-6, 1983.

42. V. I. Itin et al., "Formation of Self-Propagating High-Temperature Synthesis Products in Ti-Ni and Ti-Co Systems," _Izvestiya Vysshikh Uchebnykh Zavedenii Fizika_, No. 21, 1981, pp. 75-78.

43. V. N. Khachin et al., "Copper-Based Alloys With the Shape Memory Effect Prepared Using Self-Propagating High-Temperature Synthesis," VINITI, pp. 3370-80, 1980.

44. V. I. Itin et al., "Use of Combustion and Thermal Explosion for the Synthesis of Intermetallic Compounds and Their Alloys," _Poroshkovaya Metallurgiya_, No. 5, 1980, pp. 24-28.

45. A. G. Merzhanov et al., "Properties of WSe_2 Obtained by Self-Propagating High-Temperature Synthesis," _Izvestiya Akademii Nauk_, Vol. 13, No. 5, 1977, pp. 811-14.

46. V. K. Prokudina et al., "Synthetic Molybdenum and Tungsten Disulfides," _Poroshkovaya Metallurgiya_, No. 6, 1978, pp. 48-52.

47. A. G. Merzhanov et al., "Tungsten-Free Hard Alloy and Process for Producing Same," _U.S. Patent No. 4,443,448_, 14 February 1984.

48. Ramakrishna T. Bhatt, _Mechanical Properties of SiC Fiber-Reinforced Reaction-Bonded Si_3N_4 Composites_, NASA Technical Memorandum 87085, U.S. Army Aviation Research and Technology Activity, July 1985.

49. J. B. Holt, "Synthesis of Refractory Materials," _U.S. Patent No. 4,446,242_, 1 May 1984.

50. J. B. Holt, "Synthesis of Refractory Materials," _U.S. Patent No. 4,459,363_, 10 July 1984.

51. W. L. Frankhouser et al., _Materials Technology Underlying Space Nuclear Power Capability of the U.S.S.R._, SPC 1070, System Planning Corporation, June 1985, pp. VIII-25 through -27.

52. A. G. Merzhanov et al., "Process for Preparing Titanium Carbide," _U.S. Patent No. 4,161,512_, 17 July 1979.

53. A. P. Veselkin et al., _Fast Neutron Spectra Behind Materials and Compositions Used in Nuclear Reactor Shielding_, Atomizdat, Moscow, 1970.

54. S. L. Kharatyan et al., "Investigation of the Processes of High-Temperature Interaction of Zirconium With Hydrogen (Metals)," _Soviet Inorganic Materials_, No. 1, 1977, pp. 46-51.

55. M. E. Kost et al., "Hydrogenation of Alloys of Titanium With Scandium and Yttrium," _Russian Journal of Inorganic Chemistry_, Vol. 25, No. 3, 1980, pp. 342-45.

56. V. F. Petrunin et al., "The Position of Hydrogen Atoms in Zirconium Hydride," Fiz. Tverd. Tela, Vol. 23, No. 7, July 1981, pp. 1926-30.

57. V. V. Lunin et al., "Evaluation of Hydrogen From the Hydrides $ZrNiH_{2.88}$ and $ZrCoH_{2.88}$ in the Presence of Ethylene, Ethane, and Argon," Doklady AN. SSSR, Physical Chemistry, Vol. 266, No. 6, October 1982, pp. 1417-20.

58. S. K. Dolukhanyan et al., "Combustion of Transition Metals in Hydrogen," Doklady. Soviet Physical Chemistry, Vol. 231, No. 3, 1976, pp. 675-78.

59. S. K. Dolukhanyan et al., "Method of Obtaining Hydrides of Transition Metals," U.S.S.R. Patent No. 552,293, 1977.

60. E. V. Agababyan et al., "Combustion Mechanism of Transition Metal Under Conditions of Intense Dissociation (With Reference to the Titanium-Hydrogen System)," Soviet Combustion, Explosion, and Shock Waves, Vol. 15, No. 4, 1979, pp. 3-9.

61. A. A. Zenin et al., "Mechanism and Microkinetics of the Formation of Titanium and Zirconium Hydrides in a Synthesis Wave," Soviet Combustion, Explosion, and Shock Waves, Vol. 18, No. 4, July/August 1982, pp. 66-73.

62. W. L. Frankhouser et al., Materials Technology Underlying Space Nuclear Power Capability of the U.S.S.R., SPC 1070, System Planning Corporation, June 1985, pp. VIII-32 through -42.

63. A. G. Merzhanov et al., "Method of Producing Cast Refractory Inorganic Materials," United Kingdom Patent Specification No. 1,497,025, 5 January 1978.

64. A. G. Merzhanov, "Self-Propagating High-Temperature Synthesis," Appendix IV.5 in Advanced Materials Technology Project, SPC 1086, System Planning Corporation, June 1985.

65. W. L. Frankhouser, Advanced Materials Technology, Sections I.D. and IV.C, Status Report for December 1985 through February 1986, Contract MDA903-84-C-0225, System Planning Corporation, March 1986.

66. "Laser Plasma Nitriding and Carbiding at Metallurgy Institute," Pravda, No. 43 (24665), Cols. 2-5, 12 February 1986, p. 6.

67. V. E. Panin et al., "Explosion Compaction of the TiC-TiNi Powder Composite," Poroshkovaya Metallurgiya, No. 7, July 1985, pp. 27-31.

68. G. V. Samsonov and I. M. Vinitskii, Handbook of Refractory Compounds, IFI/Plenun, 1980.

69.   T. Ya. Kosolapova, "New Materials Based on Refractory Compounds," Soviet Powder Metallurgy and Metal Ceramics, No. 9 (201), 1980, pp. 604-14.

70.   S. M. Kats et al., "Thermal Conductivity of Lamellar Metal/Cermet Composite Materials Based on ZrO_2-Mo and Mo," Soviet Powder Metallurgy and Metal Ceramics, Vol. 24, No. 2, July 1985, pp. 146-49.

71.   G. V. Samsonov and R. A. Alfintseva, "Dispersion Strengthening of Refractory Metals--A Survey," Soviet Powder Metallurgy and Metal Ceramics, No. 2 (110), 1972, pp. 98-107.

72.   W. L. Frankhouser et al., Materials Technology Underlying Space Nuclear Power Capability of the U.S.S.R., SPC 1070, System Planning Corporation, June 1985, pp. V-18 through -20.

73.   Perspectives on Experience, The Boston Consulting Group, Second Printing, 1970.

Appendix A: SHS Program Briefing Information

(Prepared by SPC for DARPA, April 1985)

SELF-PROPAGATING HIGH-TEMPERATURE SYNTHESIS (SHS)

▲ Synthesis of compounds without external energy

Titanium Powder

+

Lampblack

Mixture

Cold Pressing

COMPACTION

SYNTHESIS IN COLD WALL VESSEL

Ignition

Combustion Wave

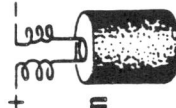

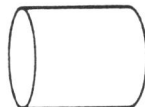

TiC Product

111

PARTICIPANTS IN DARPA'S
SHS PROGRAM

▲ Program coordination
   • Army Materials and Mechanics Research Center

▲ Lead laboratory
   • Lawrence Livermore National Laboratory

▲ Othe participants
   • Cermatec
   • Los Alamos National Laboratory
   • Monsanto Research Corporation
   • Naval Research Laboratory
   • Northwestern University
   • Rice University
   • University of California, Davis

SHS PROGRAM OBJECTIVES

▲ **Determine critical processing parameters for controlling synthesis of ceramic compounds**
  - **Reaction kinetics**
  - **Combustion mechanics**
  - **Product characterization**

▲ **Combine product densification with synthesis reactions**

▲ **Explore potential process applications in U.S. government programs**

SHS ATTRIBUTES

▲ **Process**
- High combustion temperature
- External energy not required or minimal
- Rapid heatup and cooldown
- Simple, low-cost equipment

▲ **Product**
- High chemical purity
- Varied compositions; including ceramics
- Varied forms; e.g., powders, net shapes in one-step operation, bonded laminates, gradated compositions

▲ **Overall assessment**
- Compounds, normally difficult to process, manufactured at reasonable costs

SHS REACTION PRODUCTS AND POTENTIAL APPLICATIONS

COMPOUNDS	APPLICATIONS					
	CUTTING TOOLS AND SUPERHARD ABRASIVES	HIGH TEMPERATURE STRUCTURAL	PROTECTIVE COATINGS	ELECTRICAL AND ELECTRONIC	LUBRICANTS	NUCLEAR ENERGY
• Borides	X	X	X	X		X
• Carbides	X	X	X	X		X
• Chalcogenides				X	X	
• Hydrides						X
• Intermetallic compounds		X	X			
• Nitrides	X	X	X	X		
• Silicides		X	X	X		

DARPA

SHS HAS HIGHEST REACTION TEMPERATURES AMONG INDUSTRIAL COMBUSTION PROCESSES

PROCESSES	PRODUCTS	COMBUSTION TEMPERATURES (°C)
• Combustion of hydrocarbons	Unsaturated hydrocarbons, industrial gas, carbon black	1300–1700
• Gas-flame synthesis	Oxides	1000–2500
• Oxidation treatment	Oxides	
• Blast furnace processing	Pig iron	1600–1900
• Metallothermic processing	Ferroalloys and other master alloys	2000–3000
• SHS processing	Refractory and other compounds	Up to 4000

DARPA

DARPA

POTENTIAL SHS REACTANTS ARE WIDESPREAD THROUGHOUT PERIODIC TABLE

▶ Compounds have been formed with elements designated herein

Periods

B Groups

IIIB	IVB	VB	VIB		Periods
5 11 B	6 12 C	7 14 N	8 16 O		2
13 27 Al	14 28 Si	15 31 P	16 32 S		3
	32 73 Ge		34 79 Se		4

xx yy
E

E — Element
XX — Atomic Number
YY — Approx. Atomic Weight

Transition Metal Groups

	IIIA	IVA	VA	VIA	VIIA	VIIIA	IXA	XA	IB	IIB
	21 45 Sc	22 48 Ti	23 51 V	24 52 Cr	25 55 Mn	26 56 Fe	27 59 Co	28 59 Ni		
		40 91 Zr	41 93 Nb	42 96 Mo						48 112 Cd
		72 178 Hf	73 181 Ta	74 184 W						

IIA
1 1 H
4 9 Be
12 24 Mg
20 40 Ca

Periods

Rare Earths

57 139 La	58 140 Ce	59 141 Pr	60 144 Nd

Actinide Metals

	90 232 Th		92 238 U

DEPT. OF DEFENSE

TYPICAL SHS REACTIONS

▲ Simple binary compounds
- $Ti_S + C_S \rightarrow TiC_S$
- $2Ti_S + N_{2(LG)} \rightarrow 2\ TiN_S$

▲ Compositions more complex than binaries
- $Ti_S + 0.7C_S + 0.3N_L \rightarrow Ti(CN)_S$

▲ Cermet compositions
- $wTi_S + xB_S + yC_S + zCu_S \rightarrow lTiB_{2(S)} + mTiC_S + nCu_S$

▲ Thermite-like reactions
- $3TiO_{2(S)} + 3C_S + 4Al_S \rightarrow 3TiC_S + 2Al_2O_{3(S)}$

▲ Reactions with chemical activators
- $3Ti_S + NaN_{3(S)} \rightarrow 3TiN + Na_G$

G = gas
L = liquid
S = solid

REACTION KINETICS STUDIES

▲ The mathematical statement of combustion in solids is –

$$\lambda \frac{d^2T}{dx^2} - cu\frac{dT}{dx} + \varrho Q\phi(T, \eta) = 0$$

$$\frac{ud\eta}{dx} = \phi(T, \eta)$$

$x = \infty$ $T = T_0$ $\eta = 0$ $x = +\infty$ $T = T_c$ $\eta = 1$

Where λ = Thermal conductivity Q = Heat of formation
       T = Temperature η = Fraction reacted
       c = Heat capacity ϕ = Heat generation term
       u = Wave velocity
       x = Distance
       ϱ = Density

RESULTS OF REACTION KINETICS INVESTIGATIONS

REACTION	IGNITION TEMPERATURE (°K) (MEASURED)	MELTING TEMPERATURE (°K)	COMBUSTION TEMPERATURE (°K) (MEASURED)	T_{AD} (°K) (CALCULATED)
Ti + C	2500 / 2450	1943	3004 / 2958	3210
Zr + C	2550 / 2400	2125	2832 / 3186	3400
Hf + C	2500	2500	2832	3900
Hf + 2B	2600 / 2750	2300(B) / 2500(Hf)	2873 / 2917	3520

INVESTIGATIONS OF EXTRANEOUS GAS EVOLUTION DURING SHS COMBUSTION

Gas analysis of Ti-1.5 B combustion

		Mole (Vol.) Pct.
Nitrogen	N_2	4.509
Oxygen	O_2	0.017
Carbon monoxide	CO	0.046
Hydrogen	H_2	95.28
Methane	CH_4	0.091
Ethane	C_2H_6	0.037
Propane	C_3H_8	0.012
		99.99

Total amount of gas evolved: 2.87×10^{-3} moles

Wt% of H_2O on Ti-1.5 B sample: $\sim$1%

RESULTS OF COMBUSTION
MECHANICS INVESTIGATIONS

REACTANTS	WT% DILUENT[a] TiB$_2$	COMBUSTION TEMPERATURE (°C)	WAVE VELOCITY (cm/s)
Ti-1.5 B	0	2726	1.90
Ti-1.5 B	5	2633	1.43
Ti-1.5 B	10	2605	1.28
Ti-2 B	10	2710	2.07
Ti-2 B	12.5	2645	1.50
Ti-2 B	20	2560	1.09

[a]Prereacted product material added to reactant mass

DENSIFICATION OF SHS COMBUSTION PRODUCT

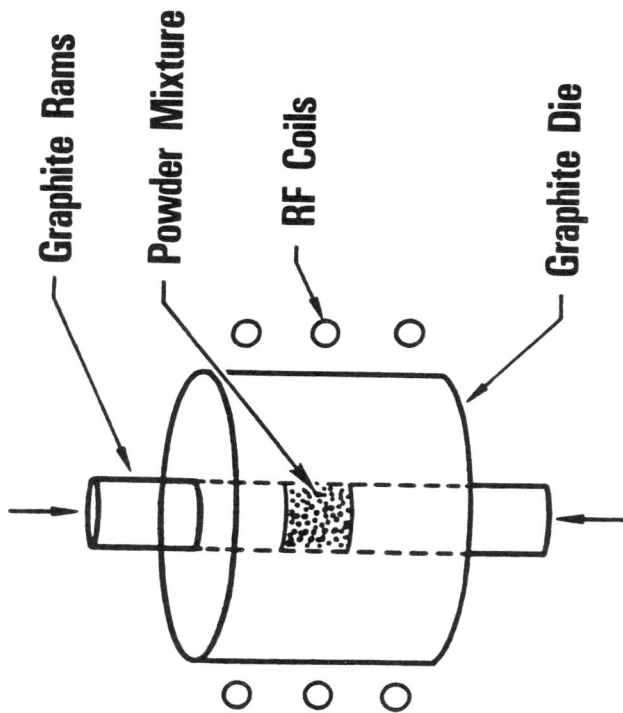

Graphite Rams

Powder Mixture

RF Coils

Graphite Die

Pressure = 4,000 psi

SHS DENSIFICATION PARAMETERS

▲ Heat source: Resistance-heated graphite die

▲ Weight of powder: 24 grams

▲ Sample dimensions (fully dense) 2.5 cm by 0.37 cm

▲ Heating rate: 1200-1400°C/min

▲ Ignition temperature: ∼1300°C

▲ Pressure: 4,000 psi

▲ Time interval for pressure application: ∼3 sec

▲ Duration of process: <2 min.

▲ Final density: 95% of theoretical

SHS PRODUCTS FORMED IN DARPA PROGRAM

▲ **Boride**
  - TiB_2

▲ **Carbide**
  - TiC

▲ **Nitrides**
  - AIN, BN, Si_3N_4, TaN

▲ β **Sialon**

▲ **Microcomposites**
  - $SiC + Al_2O_3$
  - $TiC + Al_2O_3$

SOME SOVIET APPLICATIONS OF SHS TECHNOLOGY

▲ From SHS powders
- $MoSi_2$ furnace elements
- TiC and Ti(CN) polishing pastes and abrasives
- Sulfide lubricants
- Porous filters, from Si_3N_4, TiC, TiN
- Nitrided ferroalloys, used in steel refining
- Transition metal and rare earth metal hydrides, used for radiation shielding
- Titanium nickelide, used as a "memory alloy"

▲ Dense shapes
- TiC and Ti(CN) tool bits
- (TiC + TiB_2 + Cu) "cemented" tools

▲ Weld bonding of composite structures (developmental)

SOVIET SHS POWDER
PRODUCTION, EARLY 1980s

▲ More than 200 compounds have been synthesized

▲ Pilot production reactor sizes from 2.5 to 30 liters (up to 90 kg/hr production rate)

▲ Produce more than 1,000 t/a. of:
  • $MoSi_2$
  • Si_3N_4
  • TiC

Appendix B: Self-Propagating High-Temperature Synthesis Program in the U.S.S.R.

(Prepared by SPC for LLNL, April 1985)

SOVIET SHS DEVELOPMENT HISTORY

The term self-propagating high-temperature synthesis (SHS) originated in the USSR. During experiments with solid rocket propellants, Soviet researchers decided that some exothermic reactions might be used industrially to synthesize compounds (especially ceramic and other refractory types) that were relatively intractable to fabricate by conventional techniques. The basic SHS approach was to synthesize directly from the elements. Nickel aluminide and tantalum nitride were probably the first products formed in the USSR by this approach.

The SHS concept gained political support and was organized as a national program within one of the 5-year Soviet plans of the 1970s. The initial laboratory work soon was considered to be promising enough that a pilot production facility was constructed at the lead development facility for production of powder products by the SHS process.

By the early 1980s, powder products were being produced not only in the pilot facility but also in various manufacturing facilities around the USSR. Probably TiC, TiCN, MoSi, and transition metal hydrides were produced in the greatest quantities among the many compounds that were synthesized. The Tank Production Bureau also had assumed responsibility as a major sponsor of the development program. This relationship probably was established because of extensive use of the TiC and TiCN products in machining, grinding, and polishing operations that were performed by facilities within this bureau.

During the 1970s and 1980s, many other development organizations had participated in the SHS program. Some were investigating process control and theory, and others were evaluating product materials. By the early 1980s, SHS curricula had been installed at various higher education facilities to teach the technology to engineers. Meanwhile, the primary objective in the development program had turned away from powder production to the problem of producing dense final product shapes.

128

SOVIET SHS DEVELOPMENT HISTORY

PROCESS ORIGIN -- CIRCA 1960S

- RESEARCH UNDERWAY IN SOLID-FUEL ROCKET PROPELLANTS
- POTENTIAL INDUSTRIAL PRODUCTS SYNTHESIZED (E.G., ALUMINIDES AND TANTALUM NITRIDE)

NATIONALIZATION OF DEVELOPMENT PROGRAM -- CIRCA 1970S

- NATIONAL SCIENTIFIC COUNCIL FORMED UNDER DIRECTION OF PROFESSOR A. G. MERZHANOV IN STATE COMMITEE FOR SCIENCE AND TECHNOLOGY

    CHERNOGOLOVKA BRANCH OF THE INSTITUTE OF CHEMICAL PHYSICS IN THE ACADEMY OF SCIENCES OF THE USSR BECAME LEAD LABORATORY

    PILOT PRODUCTION FACILITY FOR SHS POWDERS CONSTRUCTED AT CHERNOGOLOVKA, LATE 1970S

RECENT PROGRESS -- CIRCA 1980S

- POWDERS PRODUCED NATIONWIDE AT SEVERAL FACILITIES
- SHS TECHNOLOGY TAUGHT AT LEARNING CENTERS
- PRODUCT DENSIFICATION UNDER DEVELOPMENT AND IN PRODUCTION

SOVIET VIEWS ON SHS TECHNOLOGY AMONG INDUSTRIAL COMBUSTION PROCESSES

The Soviet Director of the SHS program always has viewed the technology as another alternative selection among industrial combustion processes. The major difference between it and the alternatives are the potentially higher combustion temperatures, as shown in the bottom entry of the chart, and the shorter combustion periods.

SOVIET VIEWS ON SHS TECHNOLOGY AMONG
INDUSTRIAL COMBUSTION PROCESSES

PROCESSES	PRODUCTS	COMBUSTION TEMPERATURES (°C)
• Combustion of hydrocarbons	Unsaturated hydrocarbons, industrial gas, carbon black	1300-1700
• Gas-flame synthesis	Oxides	1000-2500
• Oxidation treatment	Oxides	600-900
• Blast furnace processing	Pig iron	1600-1900
• Metallothermic processing	Ferroalloys and other master alloys	2000-3000
• SHS processing	Refractory and other compounds	Up to 4000

SOVIET VIEWS ON POTENTIAL SHS INDUSTRIAL APPLICATIONS

Soviet views on potential industrial applications for the SHS process almost always include the seven groups shown on the chart for simple binary product compositions. When more complex material compositions and composites are included, the list can be expanded considerably.

The specific products most often mentioned are:

- Furnace elements made from the SHS grade of $MoSi_2$ powder

- Cemented carbide tool bits (probably mostly TiC or TiC/TiB_2 or TiCN in a metallic matrix)

- Polishing pastes and grinding wheels made from TiC and TiCN powders

- Solid lubricants made from the SHS grade of transition metal sulfide powders

- Porous filters made from SHS grades of Si_3N_4, TiC, and TiN

- Nitrided ferroalloys made by SHS and used in steel melting and refining processes

- Hydrides made in massive quantities for the nuclear power industry--probably used as neutron radiation shields for terrestrial and space nuclear power systems

- Titanium nickelide memory alloys that are used to make connectors in aerospace equipment

Recently, attempts have been reported in fabricating materials that are used in microelectronic and magnetic devices. Some of the compositions reported for these experiments are titanates and zirconates; also, complex P-S-transition metal compositions have been attempted. Another potential product is magnetoabrasives. In this case, one of the compositions being investigated is TiC plus iron.

SOVIET VIEWS ON POTENTIAL
SHS INDUSTRIAL APPLICATIONS

COMPOUNDS	APPLICATIONS					
	CUTTING TOOLS AND SUPERHARD ABRASIVES	HIGH TEMPERATURE STRUCTURAL	PROTECTIVE COATINGS	ELECTRICAL AND ELECTRONIC	LUBRICANTS	NUCLEAR ENERGY
• Borides	X	X	X	X		X
• Carbides	X	X	X	X		X
• Chalcogenides				X	X	
• Hydrides		X	X			X
• Intermetallic compounds	X	X	X			
• Nitrides		X	X	X		
• Silicides		X	X	X		

SOVIET VIEWS ON SHS COMPETITIVE ADVANTAGES

The Soviet views of all potential advantages for the SHS technology when compared to alternative processes are often cited in their literature. Major credit, beyond the temperature and time factors, is usually attributed to high purity of the product, low cost, and flexibility to produce or join many new materials or combinations of materials.

In regard to chemical purity, many Soviet publications demonstrate that the product is always more pure than the starting reactants.

SOVIET VIEWS ON SHS COMPETITIVE ADVANTAGES

▲ Process
- Rapid heatup and cooldown
- High combustion temperature
- Minimal external energy requirements
- Simple, low-cost equipment
- Minimal potential for contamination

▲ Product
- Varied forms; e.g., net shapes in one-step operation, powders, bonded laminates, gradated compositions
- High chemical purity

▲ Overall assessment
- Compounds, normally difficult to process, manufactured at reasonable costs

SOVIET SHS BINARY POWDER PRODUCTS

Soviet researches claim to have synthesized more than 200 compounds by the SHS process. Most of these compounds are the simple binary type where a transition metal is combined with boron, carbon, a chalcogenide element, hydrogen, nitrogen, or silicon. In addition, two metallic elements, usually aluminum and a transition metal, sometimes are reacted together to form intermetallic compounds.

Most reactants are utilized in the solid (powder) form; however, nitrogen has been reacted as a gas or liquid and hydrogen as a gas.

SOVIET SHS BINARY POWDER PRODUCTS

► BORIDES
  - Cr, Fe, Hf, La, Mo, Nb, Ni, Ta, Ti, V, Zr

► CARBIDES
  - B, Cr, Hf, Nb, Sc, Si, Ta, Ti, W, Zr

► CHALCOGENIDES
  - Co, Fe, Mn, Ni, Ti PHOSPHIDES
  - Mo, Nb, Ta, W SELENIDES
  - Mo, Nb, W SULFIDES

► HYDRIDES
  - Nb, Nd, Pr, Sc, Ti, Zr

► INTERMETALLIC COMPOUNDS
  - CoAl, CoTi, FeAl, Nb₃Al, NbGe, NiAl, TiNi, WAl

► NITRIDES
  - Al, B, Hf, Nb, Si, Ta_C, Ta_H, Ti Zr

► SILICIDES
  - Mo, Nb, Ta, Ti Zr

Ta_C = CUBIC FORM
Ta_H = HEXAGONAL FORM

SOVIET SHS POWDER COMPOSITIONS MORE COMPLEX THAN SIMPLE BINARIES

One of the major promises in regard to potential industrial competitiveness of the SHS process is to produce compounds with complex compositions, combinations of compounds, and composites (micro-types and macrostructures). Such materials often exhibit unique properties that are not available in simple binary compounds. The SHS technology is ideally adaptable to fabricating these types of materials because extremely high reaction temperatures can be attained and reaction times can be restricted to short periods of time.

Some examples of specific Soviet accomplishments in SHS of such materials can be taken from the list in the chart to illustrate:

- Combinations of two phases (e.g., two transition metal borides or carbides) have exhibited superplastic behavior at elevated temperatures.

- Introduction of a third element within a solid solution (e.g., nitrogen into TiC) enhances fracture toughness; this is not unlike the effects of ion implantation, which is usually employed only as a surface effect.

- Direct synthesis of a cermet (a ceramic compound within a metallic matrix), as in producing the TiC/Ni tool material, has been accomplished.

- Synthesis of complex hydrides with very high hydrogen content per unit mass (e.g., zirconium-cobalt hydrides) has been demonstrated.

- Synthesis of ceramic/ceramic microcomposites (e.g., Al_2O_3 in B_4C) has been cited.

SOVIET SHS POWDER COMPOSITIONS MORE
COMPLEX THAN SIMPLE BINARIES

BORIDES
- Cr with Ti (or Zr); Mo with Ti

CARBIDES
- Ti with Cr, Sc, W; W + Mo (in various ratios)

CARBONITRIDES
- Hf, Nb, Ta, Ti (one-phase solid solutions)

HYDRIDES
- Zr with Co (or Ni)

NITRIDES
- Ti + Zr (one-phase solid solutions)

SULFIDES
- Mo + Nb, W + Nb (solid solutions)
- Combinations, with transition and rare earth metals

CEMENTED CARBIDES (cermets)
- Cr_3C_2 + Ni/Mo; TiC + Ni (or Ni/Mo); WC + Co

HETERGENEOUS MIXTURES (microcomposites)
- B_4C + Al_2O_3, SiC + Si_3N_4, TiB_2 + Al_2O_3, TiC + TiB_2, TiN + Al_2O_3, VN + Fe

SHS PRODUCTION OF TiC POWDER IN SOVIET PILOT FACILITY

The SHS powder process used at Chernogolovka in the pilot production facility is illustrated by this chart, where the product material is TiC. The reactor capacities available vary from 2.5 to 30 liters, and a block of three 16-liter reactors can average 90 kg/hr of product. Essentially continuous production is maintained by operating a block of reactors in sequential operation.

The reaction vessel is conserved for reuse by the water cooling system. The vessel walls are porous to facilitate removal of extraneous gases during the SHS reactions.

Production at the pilot facility is more than 100 tons per year for a few products. Many others are made in lesser quantities.

Soviet national SHS production of $MoSi_2$, Si_3N_4, and TiC recently reached a level of more than 1,000 tons per year for each composition.

The SHS powder production process has been available through the Soviet Licensingtorg organization for licensing outside the USSR.

SHS PRODUCTION OF T₁C POWDER
IN SOVIET PILOT FACILITY

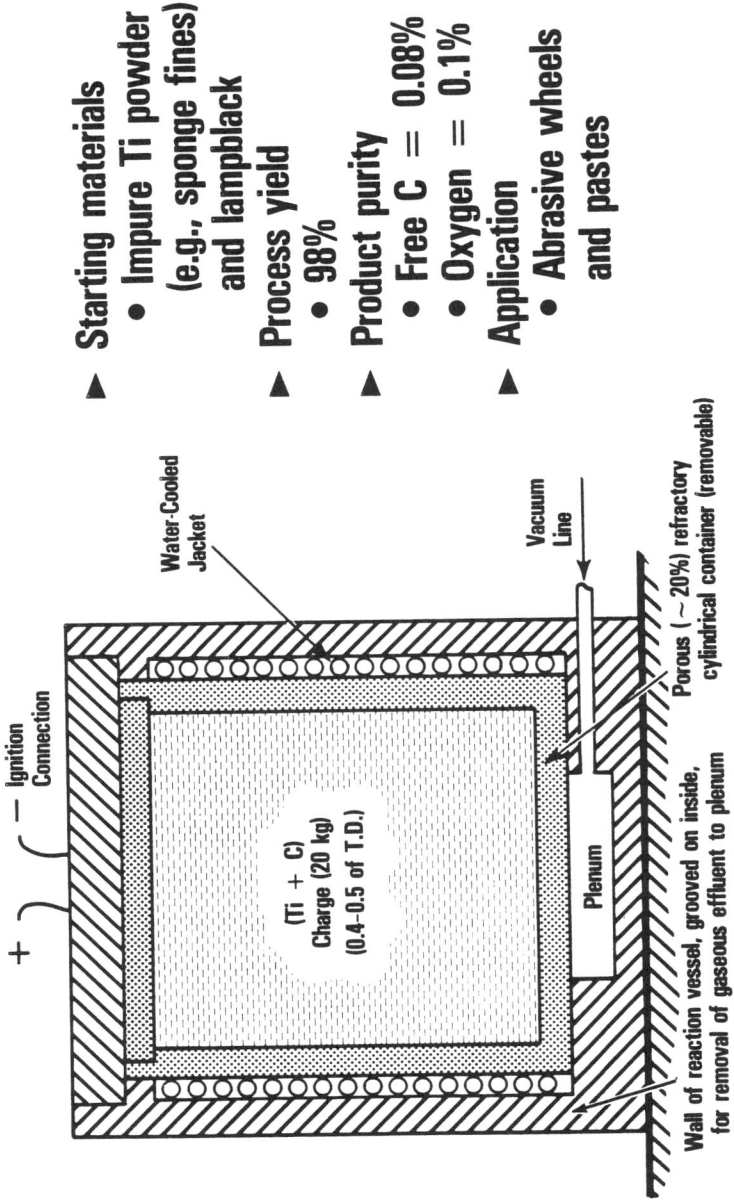

▲ **Starting materials**
  ● **Impure Ti powder (e.g., sponge fines) and lampblack**

▲ **Process yield**
  ● **98%**

▲ **Product purity**
  ● **Free C = 0.08%**
  ● **Oxygen = 0.1%**

▲ **Application**
  ● **Abrasive wheels and pastes**

Water-Cooled Jacket

Ignition Connection

+

(Ti + C) Charge (20 kg) (0.4–0.5 of T.D.)

Plenum

Vacuum Line

Porous (~20%) refractory cylindrical container (removable)

Wall of reaction vessel, grooved on inside, for removal of gaseous effluent to plenum

SOME SOVIET SHS PROCESS INNOVATIONS

During the course of the Soviet development of SHS technology, various techniques have evolved to control the synthesis reactions. Of the four such innovations listed in this chart, three are used to intensify and one to slow down the reactions. A brief description of each follows:

- Kinetic braking is used to reduce SHS reaction intensity. This is accomplished by inclusion of previously reacted product within the mixture of reactants (sometimes in portions as large as 65% of the total). This process variation has been employed in production of AlN and TiB_2.

- Thermal explosion is used to ignite some SHS reactions when a higher reaction temperature is desired. The mass of reactants is simply preheated (usually to 300°F to 600°F) until it self-ignites (and continues therefrom by self-propagation).

- Chemical furnace is a means of intensifying some SHS reactions, especially when a high mass element (e.g., W) is the metal reactant. A blanket of material that attains a more intense reaction than the desired product surrounds it and serves as a reaction booster.

- Chemical activators are added to various reactant mixtures to intensify the reactions. One example is the use of an oxide and metallic reducing agent. This technique has been employed in Soviet SHS technology to produce cast (and dense) ceramic shapes.

SOME SOVIET SHS PROCESS INNOVATIONS

▲ KINETIC BRAKING

- DILUTE REACTANT POWDERS WITH PRODUCT, E.G., TiB_2, TO REDUCE INTENSITY OF REACTION

▲ THERMAL EXPLOSION

- PREHEAT REACTANTS TO ATTAIN HIGHER REACTION TEMPERATURES

▲ CHEMICAL FURNACE

- LOW-POTENTIAL REACTANT COMPACTS, E.G., (W + C), ARE SURROUNDED BY HIGH-POTENTIAL REACTANTS, E.G., (Zr + C)

▲ CHEMICAL ACTIVATORS

- INTENSIFY REACTION BY ADDING ACTIVATORS, E.G., AZIDES, FLUORIDES, HYDRIDES, OR OXIDES WITH REDUCING AGENTS

THE SOVIET "CHEMICAL FURNACE" SHS REACTOR

The "chemical furnace" process that has been described in Soviet literature for production of WC is illustrated by this chart. In this case, compacted preforms of the tungsten and carbon mixture are surrounded by a mixture of titanium and carbon powders. The TiC reaction is ignited and it then ignites the WC reaction and boosts the WC reaction temperature.

The resultant TiC powder is easily removed by finger pressure from the surface of the reacted WC compacts.

THE SOVIET "CHEMICAL FURNACE" SHS REACTION

▲ Process

- Compact W + C to 50-60% dense
- Pre-evacuate at 10^{-2} (Hg pressure)
- React at 1-2 atm (Ar)
- Product: dense WC compacts (and crumbly TiC powder)

▲ Products claimed

- Borides — MoB_2, WB, WB_2
- Carbides — Al_4C_3, B_4C, Mo_2C, NbC, SiC, WC

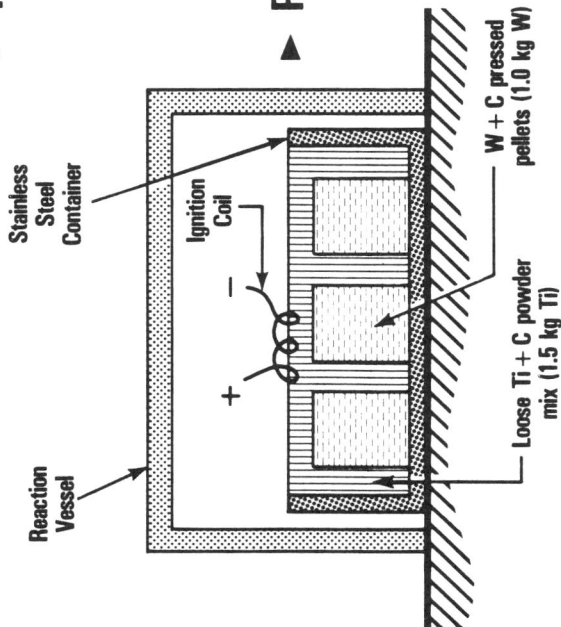

Reaction Vessel

Stainless Steel Container

Ignition Coil

W + C pressed pellets (1.0 kg W)

Loose Ti + C powder mix (1.5 kg Ti)

SOVIET PRODUCT DENSIFICATION AND BONDING COMBINED WITH SHS COMBUSTION

Although the Soviet Director of the national SHS program considers the powder production phase to have been successful, he views the production of dense, final-product shapes to be the ultimate industrial payoff for SHS technology. The three approaches to product densification that are under investigation in the USSR (and in limited production in some cases) are listed in this chart. His view on the use of SHS technology for joining composites of different materials is included on the chart (second item) in addition to product densification.

The Soviet researchers have been able to counteract the swelling tendency of SHS reactant masses by use of gas pressure in the reaction vessel. Since the greatest contributor to the swelling process is entrained extraneous gases in the reactant mass, sequential process applications of vacuum out-gassing and elevated pressure have been successful in eliminating swelling and in promoting diffusion during SHS reactions. This version of product densification is referred to as natural SHS sintering.

The Soviet view is that product density levels above approximately 90 to 95% cannot be attained through natural SHS sintering.

Highly dense products (greater than 99.8% of theoretical values) have been attained by mechanical deformation while the SHS product is still hot after passage of the combustion wave. Although extrusion pressing and rolling have been mentioned in recent Soviet literature, only pressing has been described in significant detail. In one version, mechanical pressure is applied through platens for densification; in other versions, isostatic pressure is provided through liquid or gaseous media.

Dense products also have been formed by liquefaction and casting. Liquefaction is assured by use of an oxide reactant and metal reducing agent (e.g., $3 \ TiO_2 + 4Al + 3C \rightarrow 3 \ TiC + Al_2O_3$). Densification is assured by elevated gaseous pressure or centrifugal pressure. If separation of product and oxide is desired, the centrifuge is especially useful. If a cermic/ceramic microcomposite is desired, the use of centrifugal pressure is counterproductive.

The oxide reactant promises an additional cost advantage for the SHS process since oxide powders (e.g., TiO_2) are usually considered less expensive than the elemental forms (e.g., Ti).

Soviet views about joining composite structures through SHS reactions recently have become quite positive. Dr. Merzhanov reported in 1983 that this type of application was attracting much attention.

SOVIET PRODUCT DENSIFICATION AND BONDING COMBINED
WITH SHS COMBUSTION

▲ ALTERNATIVE SHS DENSIFICATION APPROACHES

- NATURAL SINTERING (UNDER AMBIENT OR ELEVATED GAS PRESSURE) -- MAXIMUM DENSITY = 90 TO 95 PERCENT

- DEFORMATION WHILE HOT (EXTRUSION, PRESSING, ROLLING, ETC.) -- MINIMUM DENSITY = 99.8 PERCENT

- LIQUEFACTION AND CASTING (UNDER PRESSURE) -- WITH OR WITHOUT SEPARATION OF PHASES

▲ WELDING AND BONDING

o SHS REACTIONS IGNITED IN GAP BETWEEN MATERIALS TO BE JOINED

SELECTED SOVIET SHS PATENTS

A large number of patents have been obtained by USSR researchers in SHS technology. The basic concept of SHS reactions has been patented in numerous countries, including the United Kingdom, the United States, and the USSR. The basic SHS casting process has been patented in France, the United Kingdom, and the USSR; pipe casting has been patented in the United States.

As shown in the chart, other processing variants or specific synthesis reactions also have been patented. One of the most interesting items in regard to commercial potential is the U.S. patent for production of cemented carbide tool bits.

The SHS production of large quantities of hydride products also may be significant to progress in neutron shielding of nuclear reactors as power supplies in space orbit. This process has been patented in the USSR.

SELECTED SOVIET SHS PATENTS

MATERIALS PRODUCED	REMARKS, COUNTRY ISSUING PATENT
• Compounds formed with Group IV-VIII transition metals	Generic patents in USSR, USA, UK
• Compounds formed with oxide and reducing agent	Generic patent in France, UK (a casting process)
• Tantalum nitride	Use of liquid nitrogen as a reactant, USSR
• Titanium aluminide	Preheat green compacts (a "thermal explosion"), USSR
• Titanium carbide powder	Process and equipment described, USSR, USA
• Hydrides (Nd, Sm, Sc, Ti, Y)	Large batch, multi-ignition process, USSR
• Ferroalloying compounds	Nitrided by SHS reaction, USSR
• Tungsten-free cemented carbides	Densified tool bits (TiB_2 + TiC + Cu), USA
• Casting pipes or pipe liners	Two layer pipes (e.g., carbide plus oxide) formed, USA

(Prepared by SPC for DARPA, April 1985)

DYNAMIC COMPACTION
OF CERAMICS

▲ Processing approaches
- • Room Temperature Compaction — Lawrence Livermore National Laboratory
- • Hot Compaction (to 1300°C) — Battelle Columbus Laboratories

▲ Product geometries
- • Platelets
- • Cylinders
- • Tubes

▲ Product Compositions
- • Monolithics — Borides, Carbides, Nitrides
- • Composites

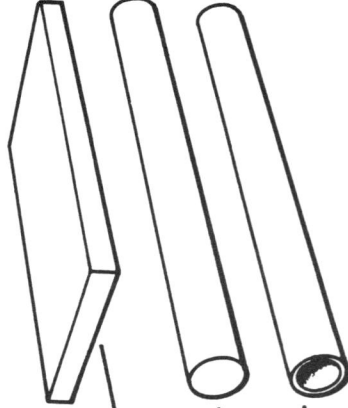

DARPA

DYNAMIC COMPACTION
OBJECTIVES

▲ Development objectives
- Determine critical processing parameters
- Demonstrate process feasibility

Explosives
Temperature

Dense Products
Oxides
Carbides
Nitrides

▲ Applicational objective
- Densify refractory ceramic products in single-step processing operation

High Density
Final Shapes

WHY DYNAMIC COMPACTION OF CERAMICS?

▲ Product purity
- Little contamination from processing environment
- Little reaction among components in composites

▲ Operational aspects
- Rapid processing rate
- Excellent reproducibility
- Minimal limitations on size scale-up

▲ Cost aspects
- Low capital investment
- Single process step to final shape

DYNAMIC COMPACTION RESULTS

Titanium Diboride
> 97% Dense
(~1" high)

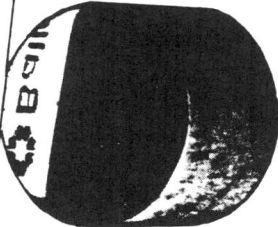

Polished Face
> 95% Dense
(~1" Dia.)

Aluminum Nitride
~90% Dense
(~3" long)

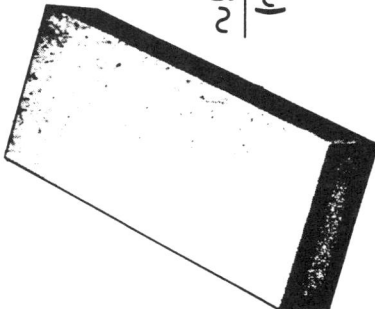

~97% Dense
(~2" O.D.)

TITANIUM DIBORIDE SOLIDS

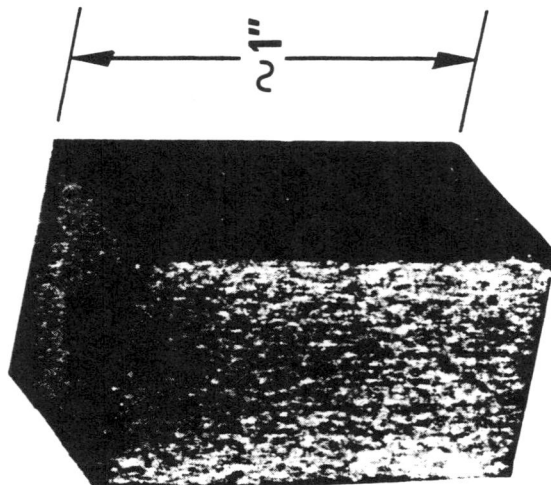

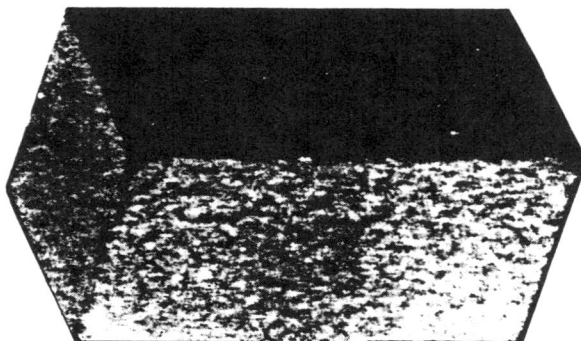

ALUMINUM NITRIDE TUBE

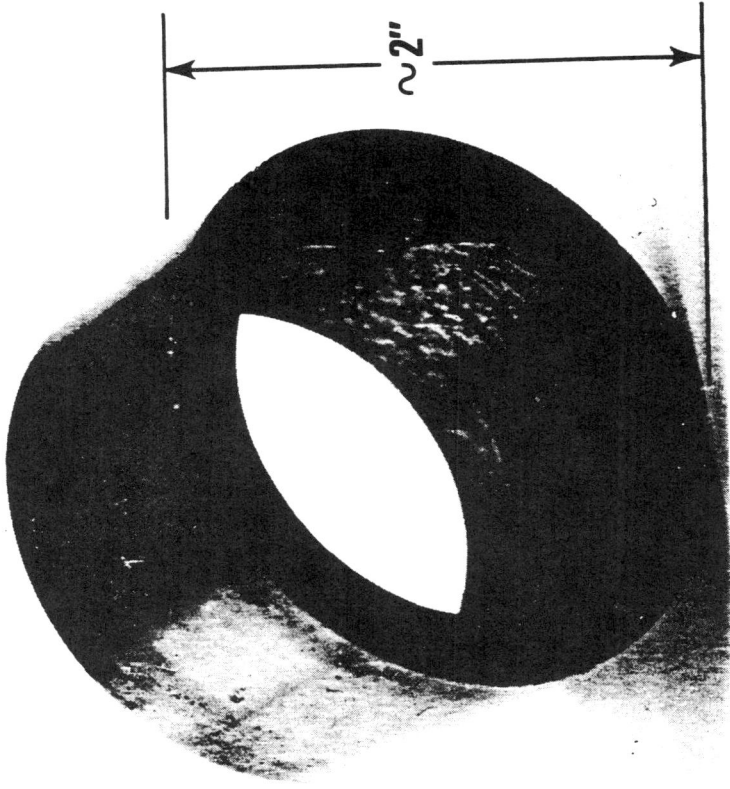

~2"

TiB$_2$ MICROSTRUCTURE AT 99% DENSITY

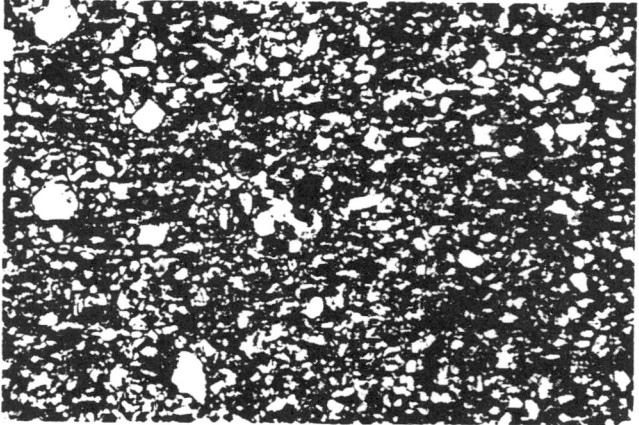

13.3 microns

50 microns

COMPACTION OF A CYLINDER

High Speed X-Ray
of Compaction Front

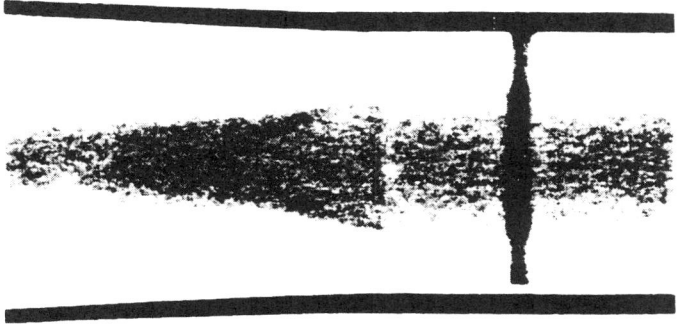

Movement of Compaction
Front

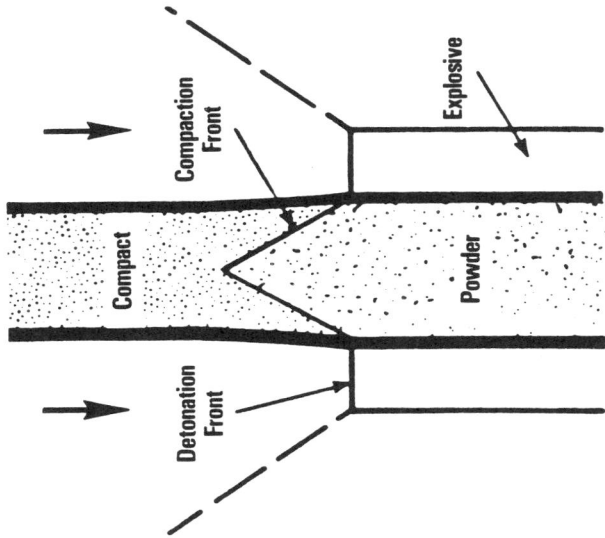

ASSEMBLY FOR COMPACTION OF FLAT PLATES

Back-to-Back Steel Cans, Loaded and Sealed

AlN Compaction Assembly
($\sim$8" × 2" × 3/8"; Each Can)

CALCULATED VELOCITY VECTORS DURING COMPACTION

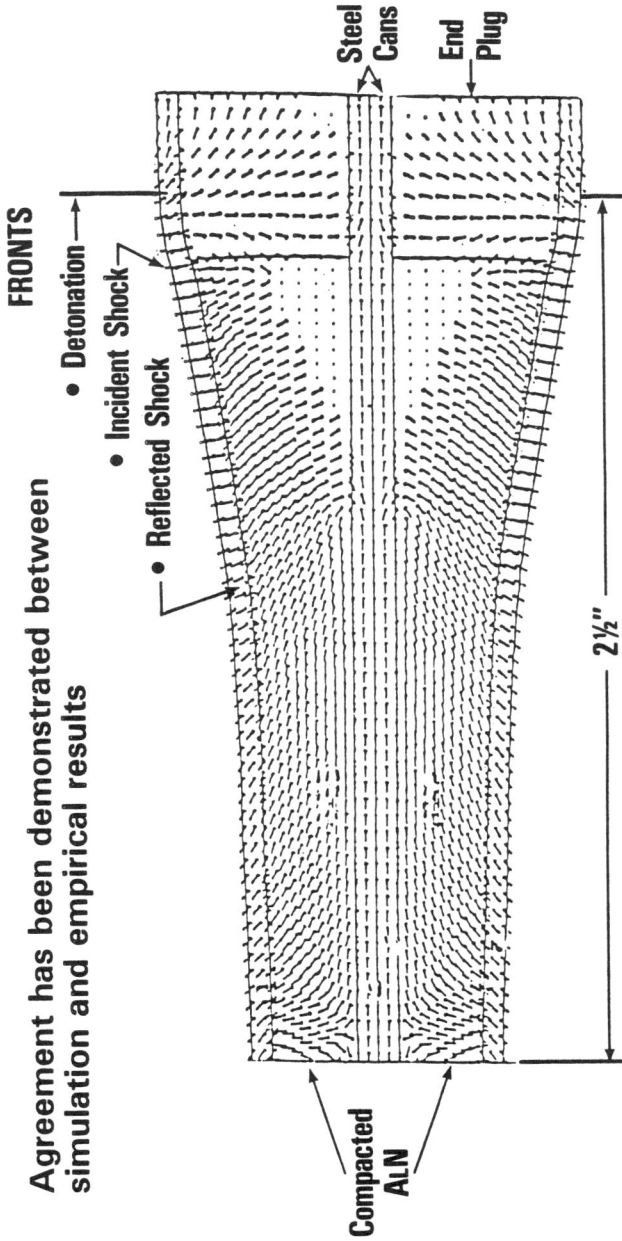

Agreement has been demonstrated between simulation and empirical results

FRONTS

• Detonation →

• Incident Shock ⌐

• Reflected Shock

Steel Cans

End Plug

Compacted AℓN

2½"

(Time = 15 μsec after detonation)

PROCESS CONTROL VS. PRODUCT PROPERTIES

AlN Properties

Detonation Velocity (km/s)	Compacted AlN in Can (Cross Section)	Density, %	Knoop Hardness
3.8		90	743
4.5		95	1046
5.6		95	983

MASS/PRESSURE/DENSITY
RELATIONSHIPS IN COLD COMPACTION

AlN Compacted to ~89% Maximum Density at R.T.

- E/M is ratio of explosive mass to compact mass
- Pressure is in GPa

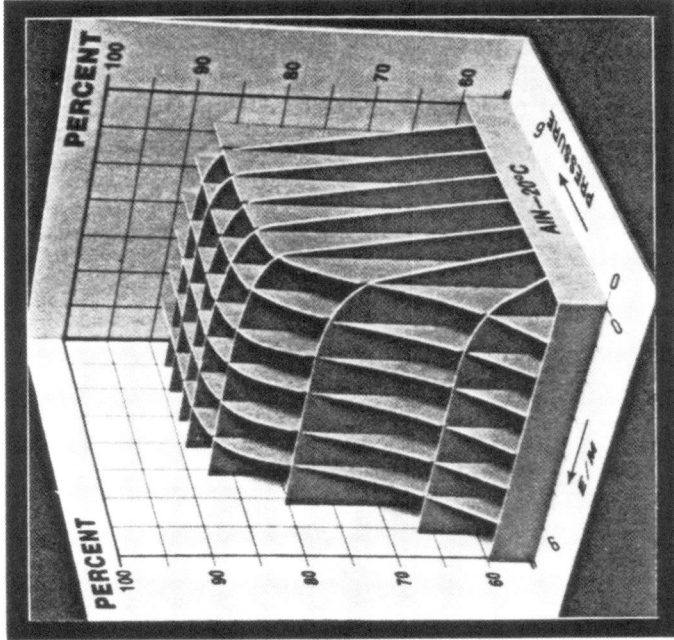

MASS/PRESSURE/DENSITY RELATIONSHIP AS INFLUENCED BY TEMPERATURE

AlN Compacted to ∼98% Maximum Density at 1100°C

- E/M is Ratio of Explosive Mass to Compact Mass
- Pressure is in GPa

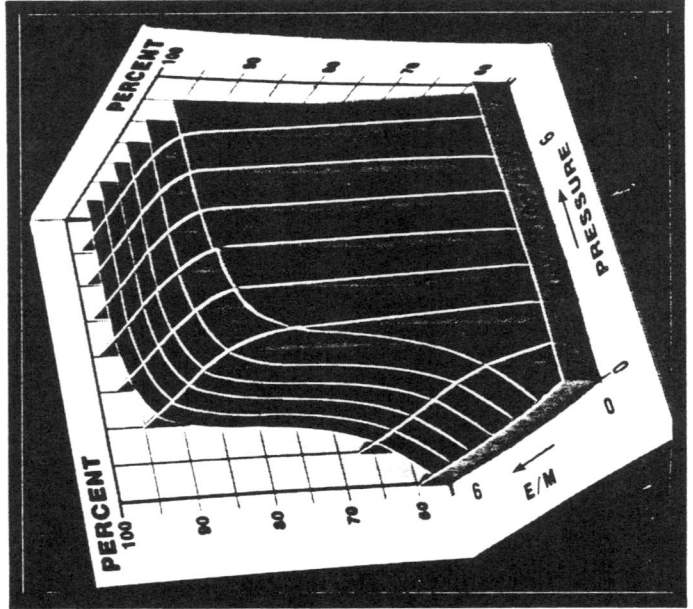

RELATIONSHIPS AMONG TEMPERATURE, SHOCK PRESSURE, AND DENSITY

ALN compacts with AMATOL explosive

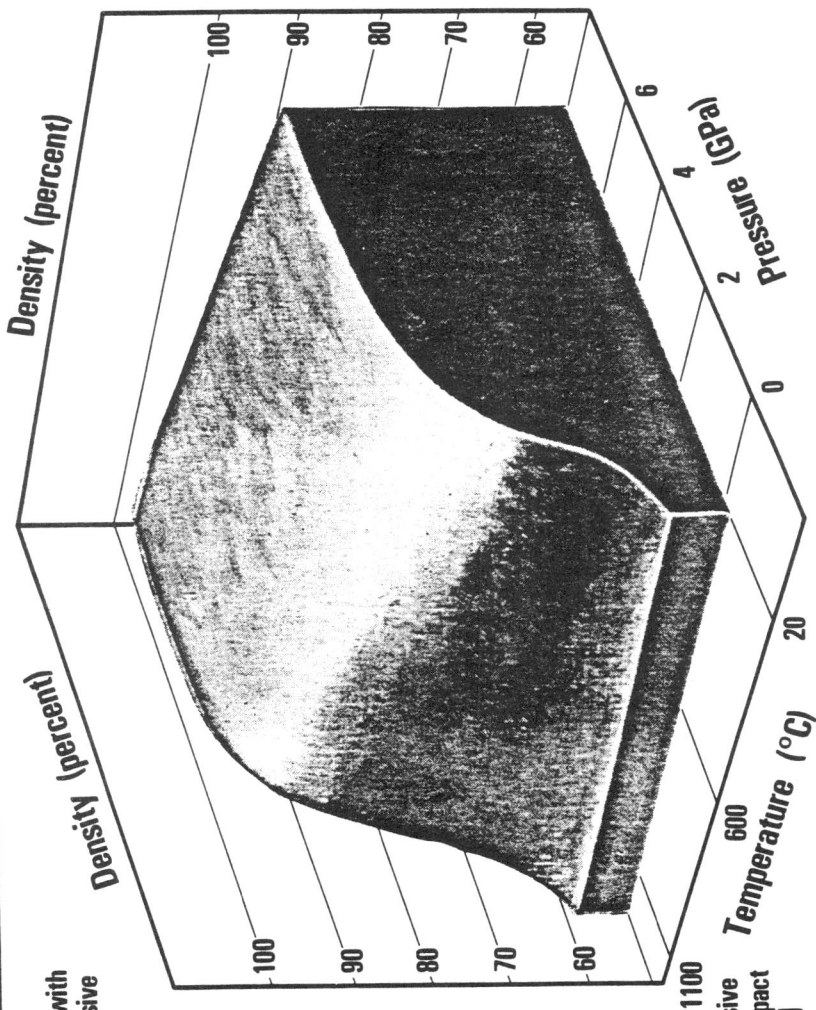

[Ratio of explosive mass over compact mass = 4]

POWDER SIZE AND COMPACTION
PRESSURE INFLUENCE COMPACTION
BEHAVIOR

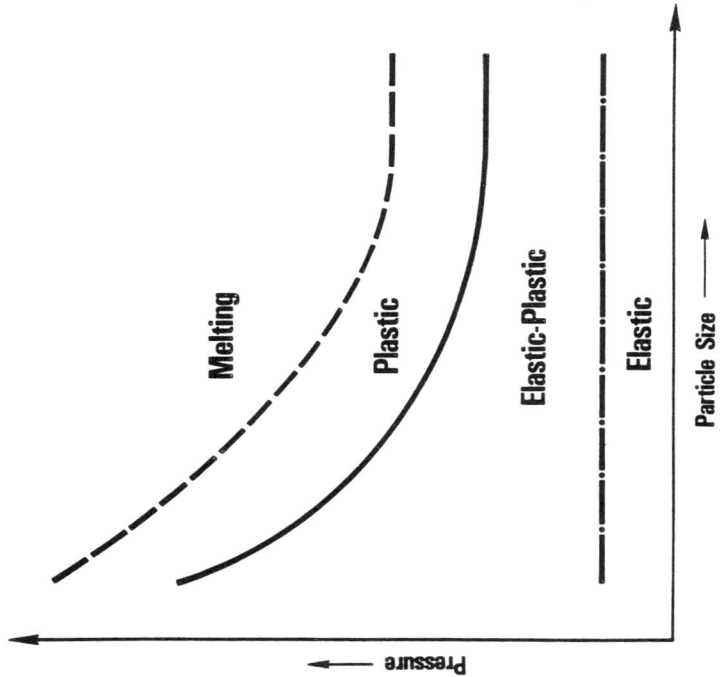

[Larger Powder
Particle Sizes and
Higher Compaction
Pressures Cause
Plastic Flow and
Melting]

- Cold Compaction
- Metallic Copper Powder

Melting

Plastic

Elastic-Plastic

Elastic

Particle Size ⟶

Pressure ⟵

INCREASED PRESSURE CAUSES MELTING

Increasing Compaction Pressure →

Evidence of Melting

Copper Particle Boundaries

~ 500X Magnification

INCREASED PARTICLE SIZE RESULTS IN MELTING

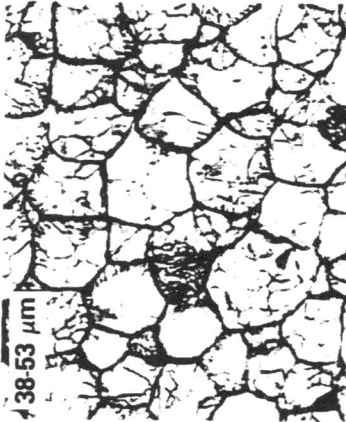

Copper Particle Sizes are Varied

106-125 μm — Plastic

1100-1700 μm — Melting

38-53 μm — Elastic-Plastic

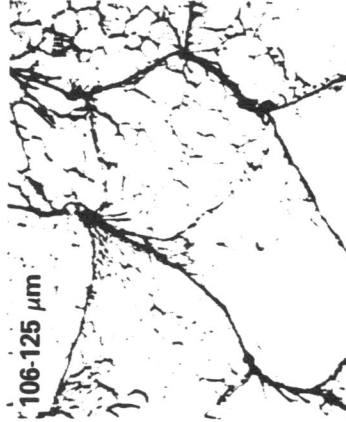

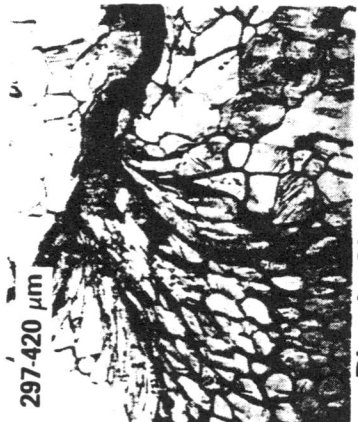

297-420 μm — Plastic Near Melting

DARPA

DYNAMIC COMPACTION
OF MACROCOMPOSITE

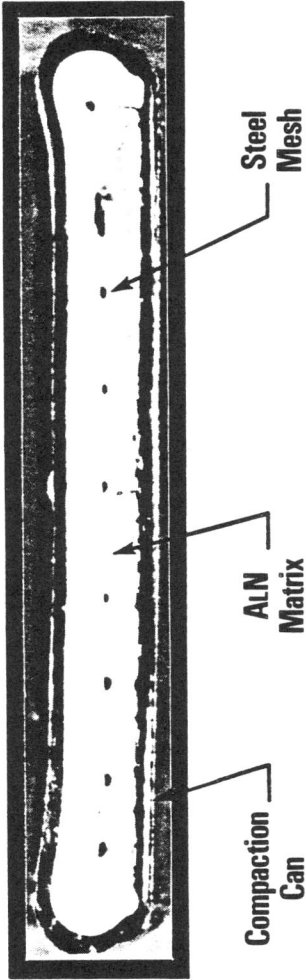

Steel
Mesh

ALN
Matrix

Compaction
Can

Steel Mesh
Compacted
within ALN

Section of
ALN after
Decanning

Appendix D: Potential Applications in Government Programs for DCT

Program Subjects	Government Office or Laboratory	Critical Products Relevant to DCP Technology Development
1. Armaments and relevant materials	Army Materials and Mechanics Research Center (U.S. Army)	Lightweight armor tiles
	Army Research Office (U.S. Army)	Ceramic gun barrels or lines
	Ballistic Research Laboratory (U.S. Army)	Duplex (metal/ceramic) kinetic energy penetrators
	Benet Weapons Laboratory, Watervleit Arsenal (U.S. Army)	Composite (metal/ceramic) periscope tubes and deep-submergence vessels
	David Taylor Naval Ship R&D Center (U.S. Navy)	
	Naval Ocean Systems Center (U.S. Navy)	
	Picatinny Arsenal (U.S. Army)	
2. Ceramic engines and components	Applied Technology Laboratory (U.S. Army)	Structural engine component (e.g., housings, liners shrouds, shields, baffles, insulators, combustion chambers)
	Lewis Research Center (NASA)	
	Naval Air Propulsion Center (U.S. Navy)	Spark plugs
	Oak Ridge National Laboratory (DOE)	Turbine blades
	Office of Basic Energy Sciences (DOE)	Roller bearings
	Office of Conservation (DOE)	

Program Subjects	Government Office or Laboratory	Critical Products Relevant to DCP Technology Development
3. Electronic and optical hardware	Lewis Research Center (NASA) Naval Research Laboratory (U.S. Navy) Office of Electronic Sciences (DARPA) Office of Naval Research (U.S. Navy)	Capacitors and semi-conductors (stacked layers) Magnets (e.g., $SmCo_5$ and metallic glasses) Integrated optics (fiber optic wave guides on ceramic substrates)
4. Other mechanical and structural components	Lewis Research Center (NASA) Los Alamos National Laboratory (SP-100 Program Office)	Nonmetallic heat exchangers and heat pipes Ceramic Cutting tools and drilling crowns Ceramic nozzles, pump seals, valves (for use in liquid metal service, turbines, weldings)
5. Rapid solidification technology	David Taylor Naval Ship R&D Center (U.S. Navy) National Bureau of Standards (DOE) Wright Patterson Air Force Base (AFWAL, U.S. Air Force)	Turbine blades, nozzles, magnets Superconductors Microelectronics

Appendix E: Soviet Engineers and Scientists Who Work in Fields Embracing the DARPA Technologies

Dynamics and Shock Waves	Self-Propagating High-Temperature Synthesis Technology	Plasmachemical Technology
G. A. Adadurov	A. P. Aldushin	M. I. Aivazov
V. P. Alekseevski	I. P. Borovinskaya	M. V. Aleeksev
A. V. Anan'in	S. K. Dolukhanyan	Ya. P. Grabis
L. V. Al'tshuler	V. P. Filipenko	G. M. Kheidemane
E. S. Atroshenko	A. K. Filoenko	V. I. Kirko
S. S. Batsanov	S. L. Grigoran	V. S. Koriagin
P. V. Belyakov	V. I. Itin	T. Ya. Kosolapova
O. N.Breusov	V. N. Khachin	C. N. Makarenko
E. Sh. Chagelishvili	S. L. Kharatyan	T. N. Miller
A. P. Cherapanov	E. I. Maksimov	S. S. Polyakov
A. A. Deribas	Yu. M. Maksimov	D. A. Prokoshkin
F. I. Dubovitskii	S. S. Mamyan	A. L. Suris
S. S. Grigorian	V. M. Maslov	V. N. Troitskii
V. G. Gorobtsov	A. G. Merzhanov[a]	A. E. Voitenko
G. I. Kanel'	Y. S. Naiborodenko	D. P. Zyatkevick
G. E. Kuz'min	A. B. Nalbandyan	
N. S. Nikiforvski	G. A. Nersisyan	
Yu. P. Raizer	S. S. Ordanyan	
O. V. Roman	A. N. Pitylin	
A. P. Rybakov	V. K. Prokudina	
A. M. Staver	V. M. Shkiro	
G. V. Stepavov	V. F. Smirvo	
Y. A. Trishin	V. I. Yukhvid	
V. M. Titov	A. A. Zenin	
F. F. Vitman		
A. E. Voitenko		
V. V. Yarosh		
Ya. B. Zel'dovich		
N. A. Zlatin		

---

[a]National Director of Soviet SHS program.

Appendix F: Soviet Journals in Materials Science, Engineering, and Applications that Were Most Useful in This Study

Soviet Title [and English Subtitle]	Areas of News Coverage
Atomnaya Energiya* [Soviet Atomic Energy]	Nuclear power reactors: design, operations, safety, materials; energy conversion
Avtomaticheskaya Svarka [Automatic Welding]	Welding, joining, and cladding processes
Detekoskopia [Soviet Journal of Nondestructive Testing]	Acoustic, magnetic, and radiation inspection and testing
Doklady AN SSSR, Khimicheskaya Tekhniki [Doklady. Chemical Technology]	Chemical process technology
Doklady AN SSSR, Khimii [Doklady. Chemistry]	Broad-based, organic chemistry and chemical reactions
Doklady AN SSSR, Physicokhimii [Doklady. Physical Chemistry]	Combustion, thermionics, materials properties
Doklady AN SSSR, Fizika [Soviet Physics. Doklady]	Mostly theoretical: aerodynamics, crystallography, cybernetics, fluid mechanics, physics, rapid solidification technology
Doklady AN SSSR, Tekhicheskaia Fizika [Doklady. Technical Physics]	Broad physics coverage
Elektronnaya Obrabotka Materialov [Applied Electrical Phenomena]	Electrotechnology in industry
Elektrotekhanika i Mekhanika [Soviet Electrical Engineering]	Industrial electrical equipment, including energy conversion
Energomashinostroenie [Soviet Energy Technology]	Energy applications in industry; power generation
Fizika Goreniya i Vzryva* [Combustion, Explosion, and Shock Waves]	Combustion, dynamic processing, fire safety, explosions
Fizika i Khimia Stekla [Soviet Journal of Glass Physics and Chemistry]	Glassmaking, materials properties
Fizika i Khimiya Obrabotka Materialov* [Physical Chemistry and Heat Treatment of Materials]	Thermal behavior of materials
Fizika Metallov i Metallovedeniye* [Physics of Metals and Metallography]	Solid state and physical metallurgy, mostly theoretical; emphasis on magnetism; rapid solidification technology included

---

*Journals that were most useful in Task 3.

Soviet Title [and English Subtitle]	Areas of News Coverage
Fizikia Tverdogo Tela [Soviet Physics. Solid State]	Solid state physics, acoustics, optics, crystallography, electro-magnetism
Fiziko-Khimicheskaya Mekhanika Materialov* [Soviet Materials Science]	Materials and properties; mechanical metallurgy
Inzhenerno-Fizicheskii Zhurnal [Journal of Engineering Physics]	Heat transfer, fluidics, electro-hydrodynamics; heat pipes
Itogi Nauki i Tekhniki: Seriia Metallurgia Tsvetnykh Metallov [Science and Technology Series on Metallurgy of Nonferrous Metals]	Nonferrous metallurgy
Izvestiya AN SSSR, Metally* [Russian Metallurgy]	Broad coverage of nonferrous metals
Khimiia Vysokikh Energii* [High Energy Chemistry]	Considerable information on plasma-chemical synthesis
Mekhanika Kompozitnyka Materialov [Mechanics of Composite Materials]	Composite materials, processing, properties
Metallofizika [Physics of Metals]	Solid state theory; plasticity of materials; metallic glasses
Metallovedeniye i Thermicheskaya Obrabotka Metallov* [Metal Science and Heat Treatment]	Ferrous and nonferrous metals: theory, processing, testing
Neorganicheski Materialy* [Inorganic Materials]	Materials processing, testing prop-erties; high-temperature materials emphasized
Ogneupory* [Refractories]	Mostly ceramic materials and processing
Poroshkovaya Metallurgiya* [Soviet Powder Metallurgy and Metal Ceramics	High-temperature materials: powder processing, properties
Prikladnaya Mathematika i Mekhanika* [Journal of Applied Mathematics and Mechanics]	Theory, measurements; mechanical and thermal physics; acoustics dynamics, ultrasonics
Prikladnaya Mekhanika [Soviet Applied Mechanics]	Structural mechanics and testing; composites; strongly theoretical

---

*Journals that were most useful in Task 3.

Soviet Title [and English Subtitle]	Areas of News Coverage
Problemy Prochnosti* [Strength of Materials]	Properties, testing, processing of materials; high-temperature materials, especially for gas turbines
Referativnii Zhurnal, Svarka [Reference Journal of Welding]	Welding equipment
Steklo i Keramika [Glass and Ceramics]	Mostly nonspecialty ceramics, (e.g., clays, porcelain); sitalls; glasses
Svarochnoye Proizvodstvo [Welding Practices]	Welding techniques and equipment
Teploenergetika [Thermal Engineering]	Heat transfer; theory, equipment design, high-temperature materials
Teplofizika Vysokikh Temperatur [High Temperature Physics]	Energy conversion, thermodynamics, heat transfer, magnetohydrodynamics; instruments, equipment, radiators
Zashchita Metallov* [Protection of Metals]	Corrosion and surface protection
Zhurnal Fizicheskoi Khimii [Russian Journal of Physical Chemistry]	Broad-based research, generally basic, in fields of chemistry
Zhurnal Neorganicheskoi Khimi [Russian Journal of Inorganic Chemistry]	Chemistry of specific inorganic materials, including molecular structures and interactions; mostly wet chemistry
Zhurnal Prikladnoi Khimii [Journal of Applied Chemistry, USSR]	Broad coverage of process chemistry, including inorganic, electro, organic, and thermodynamic
Zhurnal Prikladnoy Mekhaniki i Tekhnicheskoy Fiziki [Journal of Applied Mechanics and Technical Physics]	Materials dynamics, fluidics, sonics
Zhurnal Tekhnicheskoi Fiziki [Soviet Physics. Technical Physics]	Theory, accelerators, optoelectronics, electronics, plasmas

---

*Journals that were most useful in Task 3.

Appendix G: SPC Translations of Soviet Publications Provided in Earlier Status and Technical Reports

174

SPC Status and Technical Reports

Reporting Period	Appendix	Soviet Bibliographic Information
June-August 1985	IV-A-1	L. N. Oklei et al., "Regarding the Transition Zone in Bimetallic Specimens," *Fizika i khimiia obrabotki materialov*, No. 4, 1981, pp. 117-22.
June-August 1985	IV-A-2	V. A. Kosovich et al., "Regarding the Nature of Defects in Cylindrical Preforms Produced by Explosive Pressing of Powders," *Fizika i khimiia obrabotki materialov*, No. 1, 1982, pp. 30-34.
June-November 1985	IV-3	S. S. Batsanov "Regarding Phase Transformations and Synthesis of Inorganic Substances Under Impact Compression," *Zhurnal neorganicheskoi khimiia*, Vol. 28, Issue 11, 1983.
June-November 1985	IV-4	D. V. Fedoseev et al., "Plastic Deformation of Diamond Powders at High Pressure," *Doklady akademii nauk SSSR, tekhnicheskaia fizika*, Vol. 282, No. 3, 1985, pp. 601-604.
June-November 1985	IV-5	A. V. Kurdiumov et al., "Dynamic Recrystallization of the Sphalerilitic Modification of Boron Nitride at High Pressures and Temperatures," *Doklady akademii nauk, tekhnicheskaia fizika*, Vol. 281, No. 6, 1985, pp. 1364-66.
June-November 1985	IV-6	A. A. Artamonov et al., "Study of the Properties of the Solid Solutions of α-Al$_2$O$_3$:Cr^{3+} Produced by High-Speed Pressing of Oxide Powders," *Fizika i khimiia obrabotki materialov*, No. 4, 1984, pp. 124-30.
June-November 1985	IV-7	D. M. Karpinos et al., "Reinforced Materials Based on Magnesium Oxide," *Fizika i khimiia obrabotki materialov*, No. 5, 1983, pp. 93-95.

SPC Status and Technical Reports

Reporting Period	Appendix	Soviet Bibliographic Information

3. **Self-Propagating High-Temperature Synthesis Technology**

June-November 1984	IV-6	V. K. Prokudina et al., "Production of Aluminum Nitride of the SHS Type and of Highly Dense Ceramics Made From It," *Problemy tekhnol. goreniya; Materials of the Third All-Union conference on Technology of Combustion*, 17-20 November 1981.
June-November 1984	IV-7	Iu. M. Shulga et al., "Refractometry Measurements on Si_3N_4 Produced by the SHS Method," *Problemy tekhnol. goreniya; Materials of the Third All-Union Conference on the Technology of Combustion*, 17-20 November 1981.
June-November 1984	IV-8	S. S. Mamyan, "Study of the Possibility of Producing Boron Carbide Powder by the Self-Propagation High-Temperature Synthesis Method With Reduction Stages," *Problemy tekhnol. goreniya; Materials of the Third All-Union Conference on the Technology of Combustion*, 17-20 November 1981.
June-November 1984	IV-9	A. R. Kachin et al., "Synthesis Mechanisms and Microstructure of a Cast Hard Alloy Made From a Complex Titanium-Chromium Carbide by SHS Processes," *Problemy tekhnol. goreniya; Materials of the Third All-Union Conference on the Technology of Combustion*, 17-20 November 1981.
June-November 1984	IV-10	V. I. Yukhvid et al., "Characteristics of the Formation of Cast Tungsten Carbide in a Self-Propagating High-Temperature Synthesis Process," *Problemy tekhnol. goreniya; Materials of the Third All-Union Conference on the Technology of Combustion*, 17-20 November 1981.
June-November 1984	IV-11	A. G. Merzhanov et al., "Tungsten-Free Hard Alloy and Process for Producing Same," U.S. Patent No. 4,431,448, 20 February 1980.

SPC Status and Technical Reports

Reporting Period	Appendix	Soviet Bibliographic Information
June–November 1984	IV-12	V. I. Ratnikov et al., "Apparatus for Self-Propagating High Temperature Synthesis Processes at Extremely High Gas Pressures," *Problemy teknol. goreniya; Materials of the Third All-Union Conference on the Technology of Combustion*, 17–20 November 1981.
June–November 1984	IV-13	A. G. Merzhanov et al., "Formation of Complex Oxides With a Perovskite Structure in a Self-Propagating High-Temperature Synthesis Regime," *Problemy teknol. goreniya; Materials of the Third All-Union Conference on the Technology of Combustion*, 17–20 November 1981.
June–November 1984	IV-14	V. N. Bloshenko et al., "Degasification Under Self-Propagating High-Temperature Conditions," *Problemy teknol. goreniya; Materials of the Third All-Union Conference on the Technology of Combustion*, 17–20 November 1981.
December 1984 – February 1985	B	A. G. Merzhanov et al., "Method for Production of Two-Layer Pipe Casting," U.S. Patent No. 4,217,948, 19 August 1980.
June–August 1985	IV-B-1	A. G. Merzhanov et al., "Materials Technology Using Solid-Phase High-Temperature Reaction," *Leninskoye znamya*, No. 116 (19836), Cols. 5–7, 19 May 1985, p. 4.
June–August 1985	IV-B-2	V. K. Prokudina et al., "Study of the Effect of Starting Charge Composition on the Quality of Titanium Carbide Synthesis in Pilot-Plant Self-Propagating High-Temperature Synthesis Apparatus," *Tugoplav soedin*, 1981, pp. 29–34.

SPC Status and Technical Reports

Reporting Period	Appendix	Soviet Bibliographic Information
December 1985 – February 1986	IV-C-1	Ye. Zolotova, "Solid Combustion," *Gudok*, No. 198 (18233), 29 August 1985, p. 4, Cols. 1 and 2.
	IV-C-2	Iu. M. Maksimov et al., "High-Temperature Synthesis of the System Ti-B-Fe," *Metally (Metals)*, No. 2, 1985, pp. 219-23.

4.   Plasmachemical Technology

Reporting Period	Appendix	Soviet Bibliographic Information
June-August 1984	B-1	V. S. Koriagin et al., "Investigation of the Process of Obtaining Titanium Nitride in a Plasmachemical Reactor," *Khimiia vysokikh energii*, Vol. 7, No. 3, 1973, pp. 215-20.
	B-2	G. N. Makarenko et al., "Plasma-chemical Synthesis of Refractory Carbides," *Carbides and Alloys Based on Them*, Ukrainian Academy of Sciences, Section I, 1976, pp. 5-9.
June-November 1984	IV-1	V. V. Gustov et al., "Studies Regarding the Work of Soviet Scholars in the Field of Chemistry of High Energies," *Khimiia vysokikh energii*, Vol. 16, No. 6, 1982, pp. 490-92.
June-November 1984	IV-2	"Proceedings of International Conference on Polymerization in Plasmas and Treatment of Surfaces, Third International Symposium on Plasmachemistry," *Khimiia vysokikh energii*, Vol. 12, No. 3, 1978.
June-November 1984	IV-3	V. N. Troitskii, "Characteristics of the Physical and Chemical Properties of Powder Materials Produced by the Plasmachemical Method," *Khimiia vysokikh energii*, Vol. 13, No. 5, 1979.

SPC Status and Technical Reports

Reporting Period	Appendix	Soviet Bibliographic Information
December 1984 – May 1985	IV-1	M. Zhukov, "One Hundred Uses of the Plasmatron," *Economic Gazette*, No. 31, July 1984.
December 1984 – May 1985	IV-2	A. L. Suris, "Plasmachemical Reactors, Reactors," *Teplo-massoobmen plazmokhim. protsessakh. mater. zhdunar. shk-semin.*, Vol. 1, 1982, pp. 98-111.
June-August 1985	IV-C-2	E. A. Palchevskis et al., "Study of the Gas Phase During Synthesis of Silicon Carbide in a Nitrogen Plasma," *Izvestiia akademii nauk latviiskoi SSR, seriia khimicheskaia*, No. 6, 1981, pp. 654-57.
June-August 1985	IV-C-3	Ia. P. Grabis et al., "Formation of Silicon Oxynitride in a Low-Temperature Plasma Stream," *Izvestiia akademii nauk latviiskoi SSR, seriia khimicheskaia*, No. 6, 1981, pp. 651-54.
June-August 1985	IV-C-4	V. F. Rezvykh et al., "Effect of the the Conditions of Introducing Raw Materials on the Synthesis of Titanium Carbide," *Fizika i khimiia obrabotki materialov*, No. 2, 1983, pp. 58-61.
June-August 1985	IV-C-5	N. V. Alekseev et al., "Analysis of Physicochemical Transformations During Plasmachemical Synthesis of Titanium Carbonitride," *Izvestiia akademii nauk latviiskoi SSR, seriia khimicheskaia*, No. 1, 1984, pp. 56-60.
June-August 1985	IV-C-6	D. P. Zyatkevich et al., "Preparation of Ultrafine Boron Carbide," *Dispersionye poroshki i materialy na ikh osnove*, 1982, pp. 39-41.
June-August 1985	IV-C-7	M. Ia. Ivano et al., "Probe Temperature Measurements in the Chamber of a Plasmachemical Reactor That Is Designed for Synthesis of Ultrafine Powders," *Trudy VNII khim. reaktivov i osobo chist. khim. veshchestv*, No. 43, 1981, pp. 122-28.

SPC Status and Technical Reports

Reporting Period	Appendix	Soviet Bibliographic Information
June-August 1985	IV-C-8	Iu. G. Ionov et al., "Regarding the Methodology for Designing an Automatic Control System To Monitor Plasmachemical Processes," *Izvestiia sibirskogo otdeleniia akademii nauk SSR, seriia tekhnicheskikh nauk*, No. 13, 1983, pp. 69-76.
June-August 1985	IV-C-9	L. I. Ivanov et al., "Titanium-Ceramic Wall of a Thermonuclear Reactor," *Fizika i khimiia obrabotki materialov*, No. 5, 1983, pp. 19-21.
June-November 1985	IV-8	V. V. Averin, "Fourth All-Union Conference on Plasma Processes in the Metallurgy and Technology of Inorganic Materials," *Fizika i khimiia obrabotki materialov*, No. 3, 1984, p. 143.
June-November 1985	IV-9	V. F. Rezvyhk et al., "Some Properties of Hard Alloys Based on Titanium Carbide Produced by Plasma Synthesis," *Dispersionye, poroshki i materialy na ikh osnove*, 1982, pp. 151-54.
June-November 1985	IV-10	Yu. N. Mamontov et al., "Preparation and Study of the Structure of an Ultrafine Titanium Carbide-Molybdenum Carbide Composition," *Dispersionye, poroshki i materialy na ikh osnove*, 1982, pp. 29-32.
June-November 1985	IV-11	A. M. Bogomolov et al., "X-Ray Diffraction Study of Titanium Carbide Synthesized in a Low-Temperature Plasma," *Dispersionye poroshki i materialy na ikh osnove*, 1982, pp. 127-30.
June-November 1985	IV-12	Iu. V. Blagoveshchenskii et al., "Plasmachemical Production of Niobium and Tantalum Carbides," *Fizika i khimiia obrabotki materialov*, No. 6, 1982, pp. 32-36.

SPC Status and Technical Reports

Reporting Period	Appendix	Soviet Bibliographic Information
June-November 1985	IV-13	V. N. Troitskii et al., "The Influence of Synthesis Conditions on the Composition and Powder Size of Titanium and Vanadium Nitrides," *Fizika i khimiia obrabotki materialov*, Issue 2, 1982, pp. 24-29.
June-November 1985	IV-14	V. I. Kets et al., "Preparing Lanthanum Hexaboride by the Plasmachemical Method," *Vysokotemperatur, boridy i silitsidy*, 1982, pp. 3-7.
June-November 1985	IV-15	S. A. Malkhasyan et al., "Production Technology for Ultrafine Metal Powders Obtained by Plasmachemical Synthesis," *Teoriya i praktika poroshki metallurgii*, 1982, pp. 37-44.
June-November 1985	IV-16	A. S. Eliseeva et al., "Process Design for Plasmachemical Treatment of Atomized Solutions of Metal Nitrates," *Fizika i khimiia obrabotki materialov*, Vol. 5, 1984, pp. 47-50.
December 1985–February 1986	IV-D-1	A. A. Uglov, "Regarding the Seminar on Physics and Chemistry of Material Behavior Under Concentrated Energy," *Fizika i khimia obrabotki materialov*, No. 4, 1984.

Appendix H: Information on Soviet SHS Program that Was Released by Licensintorg

KISER RESEARCH, INC.

ITEM 1

SPHTS Trip Report
Moscow, April, 1984

1. At a meeting in the Licensintorg offices with Dr. Galchenko of the Institute of Chemical Physics, the following information was obtained in connection with the potential commercial interest of at least two U.S. companies in acquiring a license for the technology.

   1.1 <u>Productivity, physical size.</u> Reactors are available in capacities ranging from 2.5 to 30 liters. Each reactor is capable of producing the full range of powders. A block of three-eight liter reactors will produce 15 kg/hr. This group of reactors can fit into an area of four sq. meters. A block of three-sixteen liter reactors will produce 90 kg/hr. and can fit into a space six sq. meters.

   1.2 <u>Ignition cooling.</u> Ignition of the reaction is provided by the heat impulse from an electric coil. Weak exothermic reactions are carried out inside a reactor surrounded by a second reactor in which a strong exothermic reaction occurs. The heat from this reaction drives the weaker one. For example, the heat generated in the formation of TiC is used to drive the reaction to form TaC. The reactors are cooled by water jackets.

   1.3 <u>Particle size.</u> The particles should range in size from 20-50 microns. The raw material should be at least 90% pure. The Russians also claim they have the technology to use oxides as raw material to give the following result:

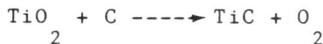

$$TiO_2 + C \longrightarrow TiC + O_2$$

   1.4 <u>Current production.</u> There are currently nine plants using the technology in the USSR. Over 1,000 T./year each of titanium carbide, moly disilicate and silicon nitride are being produced. The Russians also say they are producing combination powders of titanium carbide, tantalum carbide and combination powders cannot be made any other way, and the percent of each compound in the combination can be varied within a wide range, they claim.

P.O. BOX 33605. WASHINGTON, D. C. 20033 (202) 223.5806 TX 4991601 (ITT)

1.5 <u>Automation, reproduceability.</u> The equipment is
not automated. As the burn is carried out at such a
high temperature, it's always complete, thereby the
reproduceability of the powders approaches 100%.

1.6 <u>Patents, commercial interest.</u> U. S. patent
#3726643 with priority acknowledgement 154421 exists
as does another U. S. patent #4161512 for titanium
carbide. A Japanese company already has an agreement
for production in Japan and is negotiating for
exclusive U. S. rights, so they claim. The Russian
technology is based on a new flame theory and is in no
way related to technology covered by the Swedish patent
of Medin.

ITEM 2

SHS RESEARCH AT CHERNOGOLOVKA

Fundamental research of the Institute of Chemical Physics of the USSR Academy of Sciences on the combustion theory, in particular the study of gasless combustion of condensed systems, laid the basis for a new rapidly developing thend the self-propagating high-temperature synthesis of high-melting inorganic compounds (SHS).

Self-propagating	the velocity of spontaneous spreading of synthesis from 0,5 to 15 cm/s.
High-temperature	the temperature in the synthesis zone from 2000 to 4000°C.
Synthesis	purpose oriented production of substances and materials.

The process is based on chemical reactions which involve release of a large amount of heat and resulting in the formation of high-melting compounds of metals in directional combustion conditions.

The products of SHS may be compounds of metals of II—VIII groups of the Periodic System, such as

carbides, borides, nitrides, silicides, chalcogenides, hydrides, intermetallic compounds,

and one-phase solid solutions or well mixed heterogeneous mixtures of two or several compounds of the indicated types, or multicomponent systems containing compounds and chemical elements (solid alloys).

So far SHS method has been used to produce more than 200 different compounds, whose number is continuously increasing. Some of these are of value in terms of practical applications;

CARBIDES

TiC, ZrC (in terms of homogeneity), HfC, NbC, TaC, SiC, WC, Cr_3C_2, B_4C, ScC; TiC—WC—Mo_2C, TiC—WC, TiC—Cr_3C_2, and TiC—ScC (with various component ratios);

NITRIDES

TiN, ZrN, HfN, NbN (in terms of homogeneity), TaN_{cub}, TaN_{hex}, BN, AlN, Si_3N_4, and one-phase solid solutions of nitrogen in titanium and zirconium (in homogeneity region).

BORIDES

TiB_2, ZrB_2, HfB_2, VB_2, NbB, NbB_2, TaB, TaB_2, CrB, CrB_2, MoB, MoB_2, Mo_2B_5, WB, WB_2, WB_4, MnB, FeB, NiB, and LaB_6.

SILICIDES

$MoSi_2$, $NbSi_2$, $TaSi_2$, TiSi, $TiSi_2$, Ti_5Si_3, $ZrSi_2$.

CHALCOGENIDES

$NbSe_2$, $TaSe_2$, $MoSe_2$, WSe_2, MoS_2, WS_2, NbS_2; solid solutions: $WNbS_2$ and $MoNbS_2$.

HYDRIDES

TiH_2, ZrH_2, NbH_2, NdH_2, PrH_2, ScH_2.

INTERMETALLIC COMPOUNDS

NiAl, CoAl, WAl, FeAl, NbGe, Nb_3Al, TiNi, and CoTi.

CARBONITRIDES

(one-phase solid solutions of varying compositions)

TiC—TiN, HfC—HfN, NbC—NbN, and TaC—TaN.

HETEROGENEOUS MIXTURES

(with various component ratios)

$TiC—TiB_2$, $SiC—Si_3N_4$, $TiB_2—Al_2O_3$, $TiN—Al_2O_3$, $B_4C—Al_2O_3$, and VN—Fe.

SOLID ALLOYS

TiC—Ni, TiC—(Ni, Mo), WC—Co, Cr_3C_2—(Ni, Mo), and others.

Depending on the combustion temperature, physico-chemical properties of initial reagents and design of SHS setups, end products are available in the form of

powders of various degrees of fineness; sinters, melts, porous materials, compacts, castings.

SHS powders of high-melting compounds and materials based on them may differ in physical properties from similar compounds obtained by the conventional furnace method owing to the extremal conditions of synthesis (they exhibit higher chemical and thermal stability and abrasiveness). These powders can be used in industry to produce

> solid alloys, abrasive tools, wear-resistant and heat-proof coatings, antifriction materials, alloyed steels, refractory materials, etc.

The initial substances for synthesis are chemical elements (metals and nonmetals) or their compounds (oxides and halides) of which charge compounds are formed using a special produce. The elevated temperature developing in the reaction zone after local initiation in due to the energy resources of the initial system; that no heating device is required the considerably simplifies the technology and reduces its cost and electric energy consumption. High temperature provides for complete conversion of initial elements into end products and promotes evaporation of impurities, giving SHS products of high purity.

SHS is easy to control. By varying the combustion conditions (temperature, pressure, and reagents' ratio), it is possible to regulate the chemical and phase composition of refractory inorganic compounds obtained.

SHS meets the requirements of the revolution in science and technology the chemical processes and offers the following advantages:

> minimal electric energy requirement: low running costs; simple compact equipment; it makes possible products with controllable chemical and phace composition, various dispersion, and present porosity; allows to vary the production scale in a wide range; provides efficient environmental protection.

Please address your inquiries on licencing arrangements to Licence Department, USSR Academy of Sciences, Vavilov street, 44, korpus 2, Moscow, USSR.

ITEM 3

The development of inorganic materials possessing preset refractoriness, wear resistance, high-temperature strength and corrosion resistance usually involves great difficulties. Traditional manufacturing processes are energy-consuming and inefficient; besides, they do not always guarantee the required standard of product quality.
A new method of making inorganic materials with preset properties has been developed in the USSR. Called self-propagating high-temperature synthesis (SPHTS), it simplifies the production of such materials considerably. What is especially important, SPHTS does not need external energy sources such as plasma generators, etc.

SPHTS: A PROCESS OF PRODUCING MATERIALS WITH PRESET PROPERTIES

Prof. A.G. MERZHANOV, D.Sc. (Phys. Math.).
Head of Section. Chemical Physics Institute.
USSR Academy of Sciences

SPHTS is a special form of heterogeneous combustion with metals as "fuel" (instead of the ordinary organic fuels) and non-metallic elements (boron, carbon, nitrogen, silicon etc.) as "oxidizer" (instead of oxygen as is usually the case). With the advent of SPHTS, the chemical formula of combustion has spread to various oxygen-free processes. In the nature of its propagation, the SPHTS wave resembles the combustion of gunpowder. However practically no gases are formed in the process, and the product can retain the shape and size of the initial stock.

SPHTS reactions are accompanied by heavy heat emission (10^2-10^3 J/g), a considerable self-heating of the stock (up to 1,200-3,200 K) and bright glow. The interaction of chemical elements is localized in a zone which moves within the space taken up by the reagents at a speed of 0.5—15 cm/s owing to their heat transfer.

To date, the SPHTS method has helped to synthesize over 300 substances: borides, carbides, nitrides, silicides and other high-melting oxygen-free compounds, oxides, semiconducting compounds (sulphides, selenides, phosphides), hydrides, intermetallic compounds. Characteristically, SPHTS products have a very low content of impurities and of components that failed to enter into the reaction. The SPHTS method is also used to produce non-stoichiometric compounds with a preset concentration of vacancies, multicomponent solid solutions and metastable phases.

SHTM: NEW TUNGSTEN-FREE TOOL ALLOYS

The tungsten-free alloys produced by the SPHTS method are called SHTM (synthetic hard tool materials). They are as hard as

89-94 HRA, and their bending strength is 160-80 MPa.

Since the new method differs in principle from the conventional one, based on powder metallurgy techniques, the resultant alloys of the same composition also prove different in their structural characteristics and mechanical properties. In particular, SHTM alloys attain a certain level of hardness, with their structure being more coarser-grained than that of the alloys produced by powder metallurgy methods.

The Chemical Physics Institute has started pilot-scale production of SHTM-alloy throwaway inserts. The new tool materials have proved efficient technologically, and economically, and the process for their commercial production is under development now.

POROUS ARTICLES OF NITRIDES

If the SPHTS process is taking place in a powdered stock, the latter can be sintered to form an article of the required shape the porosity of which can be set at 1 to 50%.

This technique is used to make filters and other porous articles of titanium carbide, crucibles and electrodes of titanium nitride, a number of materials and articles of silicon nitride.

NON-POROUS ARTICLES

Various hard-alloy articles — rolls, cutting bits, dies, mould components — can also be made, employing the SPHTS method. In this case, synthesis is done in special moulds, and the product is compressed right after the combustion process is over. Materials and articles that are practically free of cores can be obtained in this way.

SPHTS's combination with other techniques of pressure shaping — extrusion, rolling, stamping, forging, impact moulding, etc — holds big promise.

The SPHTS reaction can be guided in such a way that its products will emerge in a molten state. This usually takes place when combustion involves more complicated systems, with oxides and aluminium as a reducer. Crystallization is followed by the formation of a dense material, the structure of which depends on cooling conditions and on melt composition. The range of cast materials already obtained is quite extensive — it includes inorganic compounds of various classes, all sorts of oxide systems, nickel- and chrome-base refractory alloys, mineral ceramics, including those with a melting point of 2,800—3,300 K. Producing these materials through other methods involves enormous energy expenditures, while the SPHTS process calls for almost no energy at all.

This makes it possible to use purely metallurgical methods in SPHTS, such as various kinds of mould casting — gravity, die, centrifugal. Conducted in a rotating cylindrical chamber, the SPHTS process is an effective way of producing metal pipes with a protective ceramic lining.

SPHTS: WELDING, SURFACING, IMPREGNATION

For welding, the SPHTS process is conducted in the gap between two components to be welded together, like in the case of thermite welding, with the product of SPHTS being the filler material. Using this method, articles of tungsten, molybdenum, niobium, stainless steel, graphite and other materials are now joined together in any combinations. Given the appropriate stock composition and welding conditions, seams work out stronger than the articles welded together.

This is what abrasive powders of titanium carbide, titanium diboron (1) and zirconium carbide (2) look like under a microscope. The powders have been obtained with the SPHTS method.

The SPHTS method offers a number of[2] major advantages over the induction hard-facing of worn-out components — a technique now widely-spread in mechanical engineering. A number of hard-facing compounds forming a strong and wear-resistant coating on the surface of ferrous metal articles in the process of combustion, has been developed

SPHTS melts have proved eminently suitable for impregnating porous materials, including those made by the SPHTS method. Impregnated with such melts, foundry crucibles turn out superalloy articles of superior surface quality

NITRIDING FERROALLOYS

In co-operation with the Applied Mathematics and Mechanics Research Institute in Tomsk we have developed the process of nitriding ferroalloys by the SPHTS method. The Soviet industry has brought to a commercial level the process of manufacturing nitrided ferrovanadium used as alloying composition in melting cold-strong and tool steels. Briquettes of this material, produced using the SPHTS method, are distinguished by high density, a stable high nitrogen content and ready assimilability by steel

MAKING POWDERS BY THE SPHTS METHOD

The Soviet industry is already producing more than 20 kinds of powder employing the SPHTS method. They are used for the sintering and hot pressing of articles (size fraction 0.5-3 μm), plasma and detonation spraying (40-120 μm), dynamic heterogeneous alloying of materials (40-80 μm), hard facing (over 500 μm). The SPHTS-produced powders also go into making polishing pastes and other abrasives, are used as catalysts, refractory loose protective media, in the production of

hard alloys, heat-resistant articles and coatings, antifriction ceramics, etc.

Used instead of conventional furnaces for powder making, SPHTS installations improve working conditions, release operating areas and furnace equipment, reduce energy expenditures, raise labour productivity and, in a number of cases, improve the quality of end products considerably. For instance, the production of powdered molybdenum disilicide — the basic component of high-temperature heaters — by the SPHTS method has not only reduced the manufacturing cost of the latter but enhanced their heat resistance 3-4-fold.

Owing to peculiar synthesis conditions abrasive properties of titanium carbide obtained by the SPHTS method are more strongly pronounced than those of the same material produced by the furnace or plasma method. SPHTS titanium carbide constitutes the base of new abrasive pastes, KT and KTIOL, for finishing ferrous and non-ferrous metal components. These pastes are already in commercial production.

Commercial production of silicon nitride, aluminium nitride and boron nitride powders

made by the SPHTS method, has been started.

To make powders, SPHTS products are crushed and, if necessary, classified. If a product contains undesirable substances, these are removed chemically. Powder grain size and structure depend on the stage of the process and on combustion conditions

HIGH THEORETICAL STANDARD

The SPHTS reaction was discovered in 1967 at the Chemical Physics Institute of the USSR Academy of Sciences by I.P. Borovinskaya, V.M. Shkiro and myself as an effect of self-propagation of a chemical reaction wave in mixtures of titanium, zirconium, hafnium, niobium, tantalum and other powders with boron, carbon and silicon. Borides carbides and silicides were formed as a result. It was then that this phenomenon was first used to produce nitrides as a result of metal-nitrogen interaction.

Since then, experimental methods of SPHTS process diagnosis have been developed including those of determining the character, conditions and speed of wave-front propagation, the thermal structure of the wave, the morphology, composition and structure of the products formed. We have found out the kinetic laws governing the reaction, developed thermodynamic methods of calculating product composition and combustion temperature and described the physico-chemical SPHTS models in terms of mathematics.

As a result, control techniques for SPHTS process temperature, speed and completeness and semi-empirical methods of SPHTS product composition and structure adjustment have been worked out. Chemical classes of SPHTS reactions have been broadened considerably. Besides causing metals to interact with non-metals, we have implemented SPHTS processes in metal-metal and non-metal-non-metal systems. The combustion of metals in hydrogen has been found to produce thermally unstable hydrides. The use of ferroalloys, oxides and hydrides of metals, hydrocarbons, azides and other alloys and compounds has enriched the chemistry of SPHTS and made it more varied. SPHTS processes are finding ever more varied applications in pyrotechnics, thermochemistry, preparative chemistry, metallurgy, chemistry and technology of inorganic materials